Advances in Anatomy
Embryology and Cell Biology

Vol. 63

Editors
A. Brodal, Oslo W. Hild, Galveston
J. van Limborgh, Amsterdam
R. Ortmann, Köln T.H. Schiebler, Würzburg
G. Töndury, Zürich E. Wolff, Paris

Günter H. Rager

Development of the Retinotectal Projection in the Chicken

With 66 Figures

Springer-Verlag Berlin Heidelberg GmbH

Prof. Dr. Günter H. Rager, Max-Planck-Institut für Biophysikali-
sche Chemie, Karl-Friedrich-Bonhoeffer-Institut, Abt. Neurologie, Am Fassberg, D-3400 Göttingen

As from April 1981:
Institut für Anatomie und Spezielle Embryologie der Universität,
1, Rue Gockel, CH-1700 Fribourg

This work is a revised version of the author's Habilitations-
schrift submitted to the Medical Faculty of the Georg-August
University of Göttingen, 1977.
Title of German edition: Struktur- und Funktionsentwicklung
der Retina und der retinotectalen Verbindung beim Hühnchen.
Korrelative und entwicklungsdynamische Untersuchungen

ISBN 978-3-540-10121-5   ISBN 978-3-642-67681-9 (eBook)
DOI 10.1007/978-3-642-67681-9

Library of Congress Cataloging in Publication Data. Rager, Günter,
1938– The development of the retinotectal projection in the chicken.
(Advances in anatomy, embryology, and cell biology; v. 63) A revision of
the author's habilitationsschrift, Göttingen, 1977. Bibliography: p.
Includes index. 1. Chick embryo. 2. Retina. I. Title. II. Title: The
retinotectal projection in the chicken. III. Series. QL801.E67 vol. 63
[QL959] 574.4s [598'.617] 80-15459

© Springer-Verlag Berlin Heidelberg 1980
Originally published by Springer-Verlag Berlin Heidelberg New York in 1980

Composition: Schreibsatz Service Weihrauch, Würzburg

Contents

VI

Acknowledgments

The author has great pleasure in thanking Prof. O.D. Creutzfeldt for his support, and likewise Dr. J. Houchin, Prof. H.-J. Kuhn, Dr. C. von der Malsburg, Dr. R.S. Nowakowski and Prof. R. Switzer for critically reading the manuscript, also Dr. U. Rager for the laborious work of counting and measuring optic nerve fibres, Dr. B. von Oeynhausen for his help in preparing the Golgi material, Prof. R. Switzer and Prof. Dr. I.O. Kerner for developing the best-fitting ellipse formula, Dr. B. Steffen for his advice on numerical problems, Miss Lausmann for her many years of excellent technical assistance, Mrs. Nicksch, Mrs. Pirouzmandi and Mrs. Wolff for preparing ultrathin sections, Mr. H.C. Jida for his photographic assistance, Mrs. Bast for typing the manuscript, and Zeiss, West-Germany, for making its particle-size counter (TGZ) available.

1 Introduction

Wilhelm His, one of the founders of developmental neurobiology, was convinced "that the processes of generation and development obey fundamental and simple laws and submit to the general laws of nature" (His 1901). Therefore, we should be able to find immediate conditions, dependencies and rules determining the development of an organic form. With this in mind, His (1874) defined the task of embryology as follows: "Developmental biology is essentially a physiological science; it has not only to describe how each individual form develops from the egg, it has to derive this development in such a way that each developmental stage together with all its specialities appears as a necessary consequence of the immediately preceding stage . . . Only if developmental biology has given a perfect physiological derivation for any given form, has it the right to say that it has explained this individual form." The ultimate aim of a physiological derivation would be that laws of growth valid for organic beings can be expressed as mathematical formulae (His 1874). To exemplify this, he formulated a universal and purely formal law of growth in mathematical terms making the comment: "I now suggest that the body form follows immediately from germinal growth and can be derived from the given germinal form according to the laws of growth. My interest is, therefore, firstly to detect the law of growth empirically and secondly to derive consecutive forms of the developing organism by applying this law."

We now have to ask: How can this goal be reached? First of all we have to introduce parameters and reference systems for measuring changes in position, structure and function. This would provide a reliable basis for the formulation of developmental processes as functions of time and space. The most interesting and callenging task, however, would be to find the principles, rules, and mechanisms which produce and determine the manifold appearance of these developmental processes. This would ultimately enable us to formulate developmental processes as a function of other events.

This line of research can be taken in two complementary directions. One is to investigate the genetic programme and its realization, the other is to investigate external influences which act on the genetic code and determine structural and functional development, differentiation and maturation. Since we are exclusively concerned with morphological and physiological aspects of development in the present study, we have to follow the second line, which is to start from morphological and physiological observations and to formulate functions and hypotheses that stay within the realm of morphology and physiology.

This approach has already been used by Blechschmidt (Blechschmidt 1961, 1968, 1974; Blechschmidt and Gasser 1978). He investigated morphogenetic events in locally determined metabolic fields by describing "the kinetically interdependent development of position (topogenesis), of form or shape (morphogenesis) and of inner structure (tectogenesis) of organs" (Blechschmidt 1961, p. 13). This procedure ultimately leads to a biokinetic and biodynamic understanding of development.

The importance of morphological methods is also emphasized in the theoretical work of Thom (1975, p. 154): "Of course, it is important to know how a certain protein forms . . .; but it is even more important to know *where* and *when* this protein will form and why it forms at one part of a cell while not forming in another part of the same cell." Therefore, it may be necessary first to understand determining factors in local "morphogenetic fields." Thereafter one should "attempt to integrate all the local determinisms into a coherent, stable global structure. From this point of view the fundamental problem of biology is a topological one, for topology is precisely the mathematical discipline dealing with the passage from the local to the global," (Thom 1975, p. 151). The "global form" is then defined as "the geometric figure, in space-time" (l.c. 154).

The method used in the present investigations was based on these ideas. First of all a detailed study of the normal development was made, because the questions to be asked cannot be answered by currently available data. Secondly, developmental processes were described quantitatively as well as geometrically, to provide a basis, thirdly, for investigating developmental dynamics. In a few cases it was possible to formulate developmental dynamics in terms of mathematical models which use only parameters whose values can be checked directly. Fourthly, structural and functional development were correlated in order to expand the parameter space in which normal development can be represented.

The main reasons for using the chick embryo in this investigation were (1) the ontogenetic stages of the chick are quickly and precisely reproducible, (2) although the visual system is highly developed it is relatively simple to study, because retinal fibres are completely crossed and the overwhelming majority project to the optic tectum, (3) it is much easier to operate upon and handle chick embryos than mammalian ones. Nevertheless, it might perhaps be possible by analogy to draw conclusions concerning mammalian development, since there are many similarities.

The results are presented in the following way. Section 3 is concerned with the topogenesis and morphogenesis of perikarya, dendrites, axons and synapses of ganglion cells. Their functional development is studied in Sect. 4. Morphological and physiological aspects of the development are then correlated. In Sects. 5 and 6 an attempt is made to approach developmental dynamics to explain the formation of the retinotopic map onto the optic tectum and the final maturation of ganglion cells.

2 Material and Methods

2.1 Experimental Animals

Experimental animals were chickens and chick embryos of the White Leghorn variety (Heisenberg-Nelson-Lohmann HNL) bred in our own animal house. The eggs were incubated in a forced-draught incubator (Ehret) at a temperature of 37.8° ± 0.5 °C and a relative humidity of 55%–60%. From incubation day 18 onwards the eggs were transferred to a hatching incubator running at 37.2 °C and a relative humidity of 75%–80%. Only those animals whose time of hatching was determined precisely were used for experiments in the post-hatching period. The age of all animals was measured in days and hours of ontogenesis. The embryonic stage of development was evaluated according to the Hamburger-Hamilton (HH) criteria (Hamburger and Hamilton

Incubation day	Stage (Hamburger and Hamilton 1951)
3	18
4	23
5	26
6	29
7	31
8	34
9	35
10	36
11	37
12	38
13	39
14	40
15	41
16	42
17	43
18	44
19	45
20	45
Hatching	46

Table 1. Relation between incubation days and developmental stages

1951) in order to eliminate embryos which may have been retarded. The relation between developmental stage number and day of incubation is given in Table 1.

2.2 Morphology

2.2.1 Light Microscopy

Brains were fixed with Bouin's (Romeis 1968, No. 307) or Bodian's (1937) solution and embedded in paraplast or paraffin. Serial sections were cut at a thickness of 10–12 μm. They were usually stained with haematoxylin-eosin or cresyl violet. A new reduced-silver technique was developed to stain developing axons and fibre tracts (Rager et al. 1979).

Various forms of the Golgi technique were used. The Golgi-rapid method was modified (Rager and von Oeynhausen 1979). In this modified version it gave reliable results; dendrites and axons stained equally well. The modification according to Kopsch and Colonnier (Colonnier 1964) turned out to be less suitable than the Golgi-rapid method for staining axons but it was convenient for staining dendrites. The brains of 60 embryos aged between incubation day 5 and hatching were impregnated with one of these Golgi techniques, embedded in soft Epon (Butler 1971) and cut in various section planes at a thickness of 80–100 μm. In addition, the retinae of 30 embryos were impregnated in situ with the Kopsch-Colonnier technique and later peeled off the pigment epithelium and spread onto a slide as a whole mount. However, only eight of these retinae were satisfactorily impregnated. Reproductions of Golgi-impregnated structures were reproduced either by microphotography or by drawing with the aid of a camera lucida (Leitz).

3

2.2.2 Electron Microscopy: Fixation and Embedding

In general it is quite difficult to preserve the ultrastructure of embryonic nervous tissue well. After a test series, the following fixative was found to be optimal for immersion: 3% glutaraldehyde, 1% formaldehyde (prepared from paraformaldehyde), and 0.5% acrolein in a 0.1 M sodium cacodylate buffer (pH 7.2), to which NaCl, KCl, $MgCl_2$, $CaCl_2$, and dextrose were added. Embryos older than 13 days of incubation were perfused first with a Ringer solution (20 s, 37 °C), then with the previously described fixative without acrolein (10 min, room temperature). The flow of the perfusate was roughly adjusted to cardiac output according to the formula (1/10 of body weight in grams as ml/min) given by Patterson et al (1965). The retinae were fixed exclusively by immersion. After postfixation with 2% OsO_4 in the same buffer now containing 5.4% dextrose, the tissue was dehydrated in a graded series of acetone and then embedded in an Epon-Araldite mixture modified after Mollenhauer (1964). In most cases it was stained en bloc during dehydration (0.5% uranyl acetate in 70% acetone for 2h). Semithin sections were stained either with a mixture of 1% methylene blue and 1% azure II or according to Ito and Winchester (1963). Ultrathin sections were stained with uranyl acetate and lead citrate (Reynolds 1963). Electron micrographs were taken with a Jeol 100B.

2.2.3 Localization and Orientation

The localization and orientation of all pieces of tissue embedded for electron microscopy were recorded on a chart. Thirty six eyes of embryos aged between 3 and 20 days of incubation were fixed with Bouin's solution and embedded in paraplast to enable localization of degenerating cells and to measure the growth of the retina during this period of development. All these retinae were cut perpendicularly to the pecten. Sections whose plane was near the lens centre were examined with oil immersion in a light microscope, the stage controls of which were connected to a plotter via linear potentiometers. Each degenerating cell was marked with a dot within the outline of the eye cross-section. The length of the strip of retina seen in this central plane was measured with the semiautomatic image analysis system ASM of Leitz.

The retinal papilla was left attached to the nerve to provide a marker for orientating the optic nerve at the various developmental stages. The angle between the pecten axis and the vertical meridian was measured in embryos of incubation day 18 and in newly hatched chicks; a stereotactic device was used to keep the head in the normal horizontal plane (defined in Sect. 2.3). The angle is about 30°, approximately the same as in the pigeon retina (Galifret 1968). The retina and the cross-section of the optic nerve were divided into standard quadrants (McGill et al. 1966) with the pecten as a reference.

In all cases a cross-section of the optic nerve was obtained from the region where it leaves the orbit and enters the skull cavity. In this region the cross-section is approximately circular. The entire cross-section could be placed in the electron microscope up to the stage of hatching. Nerve cross-sections of older animals were larger and were divided into quadrants by cutting them with a glass knife in steps of 50 μm on an ultramicrotome (OMU III Reichert). Thus, changes of the nerve's cross-sectional area were kept minimal. At each 50 μm step semithin sections were made for control. The changes of the cross-sectional area did not exceed 1%.

2.2.4 Measuring Procedures

The total number of optic nerve fibres at any developmental stage was estimated by measuring the cross-sectional area of the nerve and multiplying this area by the mean fibre density. Both measurements were taken from the same ultrathin sections in order to obtain reliable values for the number of fibres per nerve. Thus, although the area of the ultrathin section was up to 20% smaller than the corresponding semithin section because of compression effects, the estimate of the total counts was not affected. The total area of the ultrathin section was measured by following the pial border of the nerve anticlockwise and recording the X-Y-coordinates of on average 30 to 40 points on the circumference. The circumference together with these special points was traced out by means of a plotter connected to the stage control of the electron microscope with linear potentiometers for an overall check. The coordinates of these points were fed into a computer. For the small cross-sections a polygon was constructed (Fig. 1). Since nerve cross-sections of older animals had to be divided into quadrants, the area of the whole cross-section was estimated from the best-fitting ellipse found by the method of Gauss; the perpendicular distance between the measured points and the ellipse first found by assuming initial values is minimized by an iterative process (Fig. 2). The cross-sectional area of the ellipse was compared with that obtained by the polygon computation at several embryonic stages. It was found that the difference between these two approximations was always less than 1%. Each procedure was used twice to measure the area of the ultrathin section, the section being in a different position in the electron microscope each time. The average error amounts to less than 2%.

In addition, a rectangle was drawn around the nerve so that each side made a tangent to the nerve's circumference. The rectangle was subdivided into small fields each of which was the size of an electron micrograph. This set of adjoining fields was the population from which samples were selected, either systematically in a lattice-like arrangement or randomly. The orientation of the rectangle relative to the nerve cross-section was arbitrarily chosen so that even in the case of the lattice pattern a random sampling was guaranteed. Thus each small field had the same chance of being included in the sample. In a test series with 240 micrographs it was found that at least 60 micrographs were needed for a good estimate of the population density, i.e. for the standard error of the mean $(s_{\bar{x}}/\bar{x})$ to be less than 5%. This meant that, on average, 10 000 fibres were counted per nerve. The mean and the standard deviation of the fibre density were determined and the density distribution was tested for normality with the chi-squared test. To get the total number of nerve fibres, the mean fibre density was multiplied by the mean cross-sectional area. The final errors occurring in the measurement of fibre density and of the cross-sectional area of the nerve were estimated from the component errors by summing according to Gauss' error law.

The total magnification used was × 35 000 for embryonic stages and × 24 500 for hatched animals, to facilitate identification of small axons and glial tongues. After each series of micrographs, a grid was photographed to calibrate the pictures.

As mentioned already, each cross-section could be subdivided into standard quadrants. In addition, the computation of a polygon or a best-fitting ellipse allowed demarcation of a central core and of a peripheral ring (Figs. 1, 2). Thus, fibre density, fibre diameters, percentage of myelination, and the volume fraction of tissue compartments could be evaluated separately in each of these regions.

5

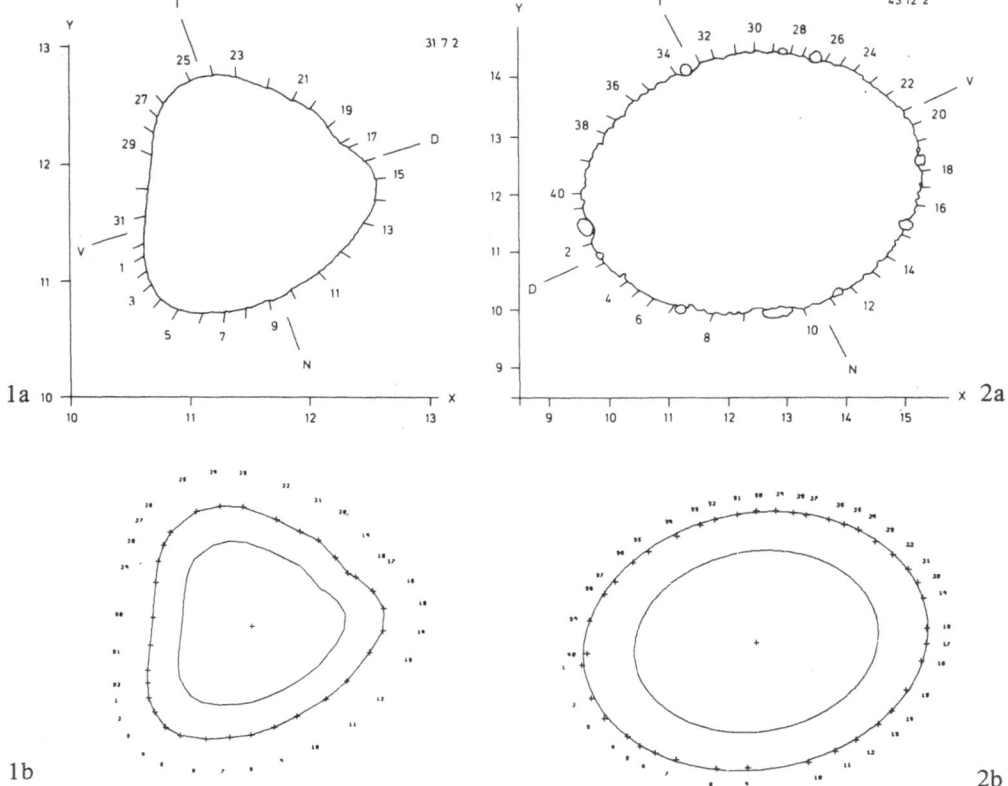

Fig. 1a, b. A polygon approximation to an optic nerve cross-section taken from a 7-day-old embryo. The upper figure shows the pial border of the nerve. It was followed anticlockwise and recorded on a plotter which was connected to the stage control of the electron microscope via linear potentiometers. The nerve is divided into standard quadrants (*D*: dorsal; *N*: nasal; *V*: ventral; *T*: temporal). Readings from the *X*- and *Y*-micrometers of the stage controls are given (a). The actual linear movement of the stage was only 1/6 of this value. The coordinates of 32 points on the pial border were fed into a computer and a polygon was constructed (b). The central cross represents the centre of gravity of the polygon. The inner polygon concentric with the pial border demarcates the central core of the nerve; by definition its area is half that of the whole cross section

Fig. 2a, b. Approximation to an optic nerve cross-section from a 17-day-old embryo with a best-fitting ellipse. The conventions are the same as those in Fig. 1

Fibres were usually identified according to well-known criteria (Vaughn and Peters 1968; Hughes and Wässle 1976; Vaney and Hughes 1976) in the fibre counts. These criteria, however, were not sufficient for identification of axons at early embryonic stages. As is shown in Fig. 3, glial processes also contain microtubules and microfilaments. Thus, in many cases additional criteria such as the presence and shape of mitochondria, the cytoplasmic background, and the packing density of microtubules were used.

Fibre diameters were measured by superimposing circles of various diameters on the fibre image. Later on, a particle-size counter TGZ (Zeiss) was used. Fibre diam-

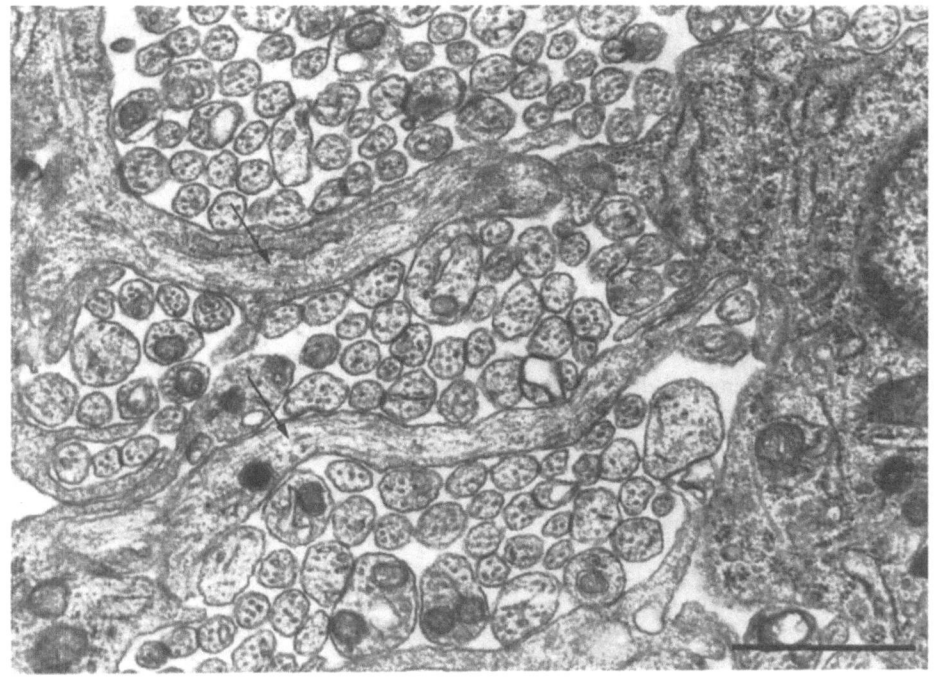

Fig. 3. Optic nerve of an 18-day-old embryo. Glial processes (*arrows*) often contain microtubules and microfilaments. *Bar* 1 μm

eters could only be measured to within 0.25 μm because of the unavoidable variability caused by fixation and embedding processes.

The proportions of glial cells, blood vessels, and fibres and their changes throughout development were measured by a systematic point count. For this purpose a square lattice of about 400 points was computed and placed randomly over the whole cross-section (Hilliard and Cahn 1961; Hally 1964; Hilliard 1968). The volume fraction and the distribution of the various compartments in the nerve were determined by identifying the structure lying at each lattice point. The volume fraction $V(V_A)$ of each compartment was calculated from the formula

$$V(V_A) = \frac{P_A}{P} \tag{1}$$

where P_A is the number of lattice points falling on component A and P is the total number of points falling on the nerve tissue.

To optimize the fibre-pathway function (Eq. 21) in the competition model, it was necessary to determine the time required for fibres to travel from the site of the optic nerve cross-section to the tectal region in which fibres terminate. Tissue shrinks variably with age and fixative (Leibnitz 1967), so the length of the fibre pathway at each developmental stage was determined physiologically in vivo by measuring the distance between stimulating (optic nerve papilla) and recording electrodes (central area of the optic tectum) (Rager 1976a).

2.2.5 Optimization of Parameter Values

The fibre-diameter function (Eq. 15), the fibre-pathway function (Eq. 21) and the generation-degeneration function (Eq. 26) resulting from the respective models were fitted to the measured values. Since these functions were non-linear, the Gauss-Newton method was used (Hartley 1961; Peil 1970; Scharf 1971; Scharf and Peil 1975); this method minimizes the sum of squares of the differences

$$Q = \Sigma \, (y_i - f(x_i))^2 \rightarrow \text{Min} \tag{2}$$

between measured values y_i $(i = 1, \ldots, N)$ and values obtained from a function $f(x)$ at the position x_i. This minimum is reached by forming the partial derivatives of Q with respect to the parameters a_k $(k = 1, \ldots, K)$ which have to be optimized. These derivatives should be equal to zero:

$$\frac{\partial Q}{\partial a_k} = 0.$$

The function $f(x_i, a_k)$ is expanded in a Taylor series

$$f(x, a_1, \ldots, a_k) \approx f(x, a_1^{(0)}, \ldots, a_k^{(0)}) + \sum_{k=1}^{K} h_k \left(\frac{\partial f}{\partial a_k}\right)^{(0)} \tag{3}$$

where the non-linear terms may be neglected (Zurmühl 1965; Peil 1970; Scharf and Peil 1975). The variable h_k is the increment in parameter a_k; $\partial f/\partial a_k$ is the partial derivative with respect to parameter a_k; and $a_k^{(0)}$ is the initial value of parameter a_k. By replacing $y_i - f(x_i)$ by ϵ_i we obtain the linear function Q

$$Q = \sum_{i=1}^{N} \left(\sum_{k=1}^{K} h_k \left(\frac{\partial f}{\partial a_k}\right)_i^{(0)} - \epsilon_i\right)^2 \rightarrow \text{Min} \tag{4}$$

It can be minimized by forming partial derivatives with respect to the increments h_k

$$\frac{\partial Q}{\partial h_k} = 2 \times \sum_{i=1}^{N} \left(\sum_{k=1}^{K} h_k \left(\frac{\partial f}{\partial a_k}\right)_i - \epsilon_i\right) \times \left(\frac{\partial f}{\partial a_k}\right)_i = 0 \tag{5}$$

This is a system of normal equations (5), from which h_k can be computed. A new initial value $a_k^{(1)}$ is defined in terms of the initial values $a_k^{(0)}$ and the increment h_k multiplied by a factor v $(0 \leqslant v \leqslant 1)$

$$a_k^{(1)} = a_k^{(0)} + v \times h_k^{(1)} \tag{6}$$

and used in the next iteration. The factor v was introduced by Hartley (1961) to guarantee the convergence of the Gauss-Newton method. Q (Eqs. 2,4) is evaluated for $v = 0$, $v = 0.5$ and $v = 1$. The three points $Q(0)$, $Q(0.5)$ and $Q(1)$ define a parabola, whose minimum is obtained from

$$v_{\min} = 0.5 + \frac{(Q(0) - Q(1))}{4 \times (Q(1) - 2Q(0.5) + Q(0))} \tag{7}$$

The iterative procedure is stopped, when the difference between the residual errors S_R of the nth and the $(n + 1)$th iteration is smaller than a pre-selected value ϵ. The residual error is defined as

$$S_R = \sqrt{\frac{\sum\limits_{i=1}^{N} [y_i - f(x_i)]^2}{N - K}} \tag{8}$$

Satisfactory initial values for the generation function (Eq. 20a) were found by a logit regression (Wingert 1969; Kretschmann and Wingert 1971), those for the fibre-diameter function (Eq. 15) and fibre-pathway function (Eq. 21) were obtained with the method of internal least squares (Hartley 1948; Scharf 1969, 1970; Scharf and Hoffmann 1971). None of these methods could be applied for the degeneration function (Eq. 25). Therefore the ranges of the parameters α, k_3, s_0 were estimated and a regular lattice was constructed in this parameter space. The error function was computed for each lattice point. The Gauss-Newton iteration was started with those values which gave the least squares. The fibre-pathway function was optimized first, then the optimal parameters for the generation-degeneration function (Eq. 26) were computed. The coefficient of determination derived from the Pearson product-moment correlation (Scharf 1971) was used to measure the fit of these functions. The standard deviations of the optimized parameters were determined according to Zurmühl (1965).

Since some of the observations were more precise than others, they should influence the approximation more strongly than less precise ones. Thus, the measured values were weighted with the factor g_i which was inversely proportional to their variance. The sum of all g_i was always equal to 1 (van der Waerden 1971).

A UNIVAC 1108 (Gesellschaft für wissenschaftliche Datenverarbeitung, Göttingen), a PDP-11 and a HP 9830 desk-top computer were used for the computations.

2.3 Physiology

In order to keep the embryos in optimal condition a chamber was constructed in which temperature and humidity could be controlled. The eggs were opened at their blunt pole, the membranes were slit carefully in order to avoid bleeding and the head was fixed in the stereotactic device in such way that the horizontal plane was defined by the interaural line being 1 mm above the ventral rim of the upper beak at the rostral margin of the external nares (Miles 1972). Since the heads of embryos younger than 14 days of incubation were too soft to be held by ear bars, special plastic moulds were made which held their heads without causing injury and gave the same coordinate system as used in older animals.

Before incubation day 17 the embryos were curarized with a mixture of gallamin (Flaxedil, Boehringer) (20 μg/g/h) and d-tubocurarin (Curarin-Asta) (1.5 μg/g/h) (Miles 1972) dissolved in Ringer solution. Older animals were first anaesthetized with Equithesin (3–5 μl/g) (a mixture of sodium pentobarbital, chloral hydrate, magnesium sulphate, and a preservative), or with metodimate hydrochloride (Hypnodil, Janssen) (approximately 10 μg/g), or with urethane (1.6 mg/g). When single cells had to be recorded, those animals which had started breathing (incubation day 19) were artificially respirated through the left thoracic air sac with a mixture of 95% O_2 and

5% CO_2. The rate of flow was measured by a flowmeter. A valve protected the animals against excessive pressure. The body temperature was kept between 37° and 38 °C. The electrocardiogram was always taken as a monitor.

The left eye was opened near the ora serrata, and the stimulating electrode inserted through the vitreous humour in order to stimulate the optic nerve head. The precise position of the electrodes and the duration, strength, and polarity of the stimulus were adjusted so that a maximal response was obtained. Recordings from the tectal surface were made either with chlorided silver ball electrodes or with micropipettes. The reference electrode was positioned either in the beak lumen or in the neck muscles.

The central tectal area was selected because it is most advanced in development. It is situated ventrolaterally in the rostral third of the tectum (LaVail and Cowan 1971a; Crossland et al. 1974). The micropipettes were inserted perpendicular to the surface and traversed the tectum in a radial direction. For optic tract recordings micropipettes were introduced from above on stereotactic coordinates. The micropipettes were filled with 2 M sodium chloride or 1.5 M potassium citrate (resistance between 8 and 20 MΩ) or with a dye (Niagara Sky Blue). The silver ball electrodes were coupled to capacitors (bandwidth 1.6 Hz–3 kHz) or directly (bandwidth 0–3 kHz). The signals from the micropipette were fed directly into a field-effect transistor

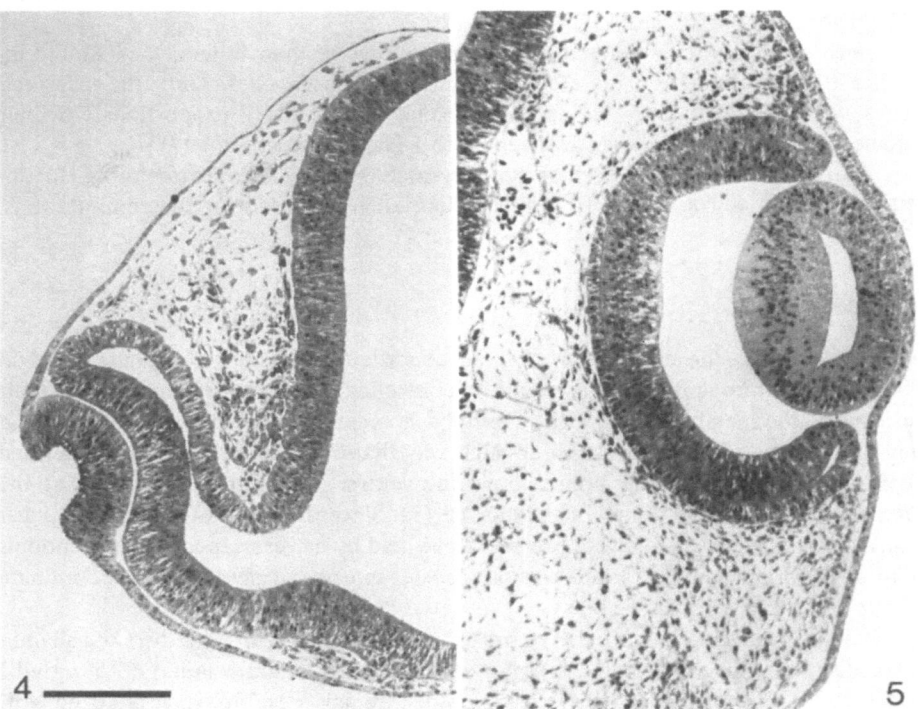

4 ———

5

Fig. 4. Frontal semithin section through the head of an embryo at stage 15. The formation of the optic cup has begun. The lens epithelium has already caved in. *Bar* 150 μm

Fig. 5. Eye of an embryo at stage 18 (horizontal semithin section). Retina and pigment epithelium are adjacent to each other except for small regions at the optic cup's margin. The lens now has a vesicular form and is covered with epithelium. *Bar* 150 μm

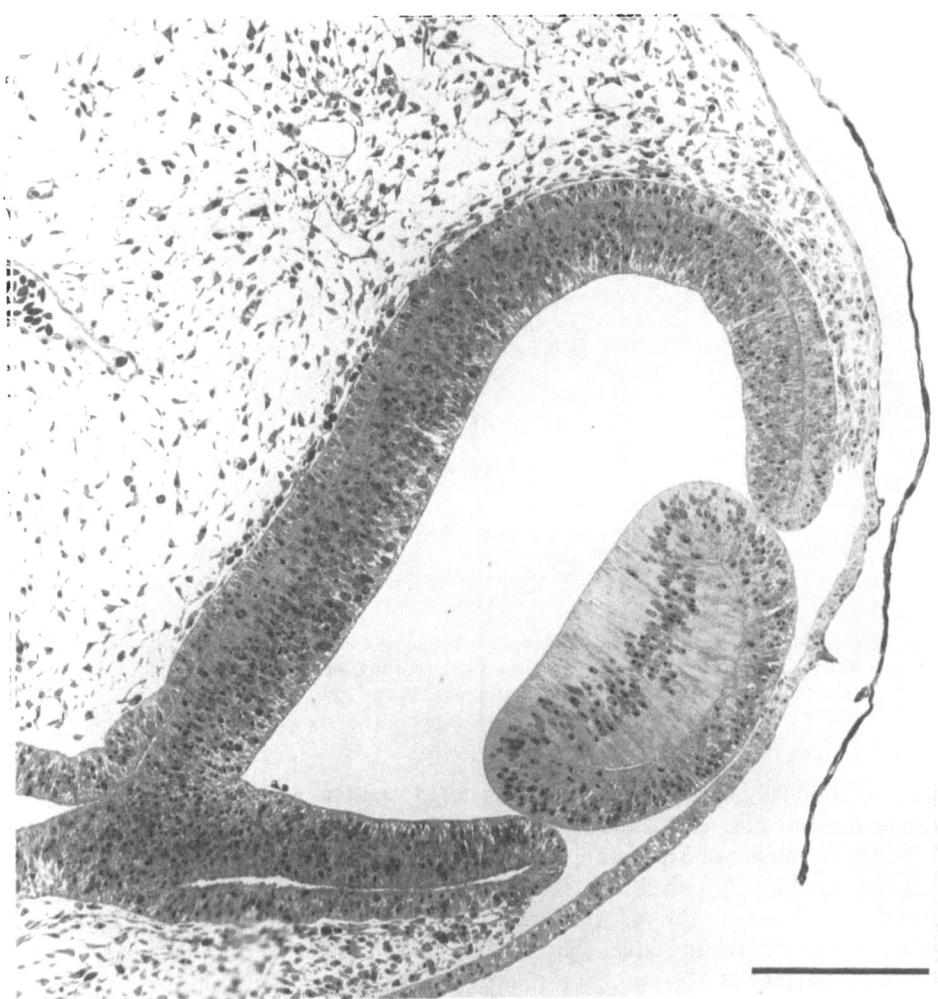

Fig. 6. Frontal semithin section through the eye at stage 20. Central retina and pigment epithelium are curved towards the lens. The choroid starts to form in the same region. *Bar* 150 μm

source-follower. The preamplifier had a capacity negative feedback. The potential displays on the oscilloscope were filmed. Data analysis was carried out on a CAT 1000 or on a PDP-11 computer. The distance between stimulating and recording electrodes was calculated from stereotactic coordinates X, Y, and Z.

3 Topogenesis and Morphogenesis

3.1 Ganglion Cell Perikarya

The optic vesicle is originally convex, but it becomes invaginated at stage 14 (after approximately 50 hours of incubation) and forms the optic cup whose future inner

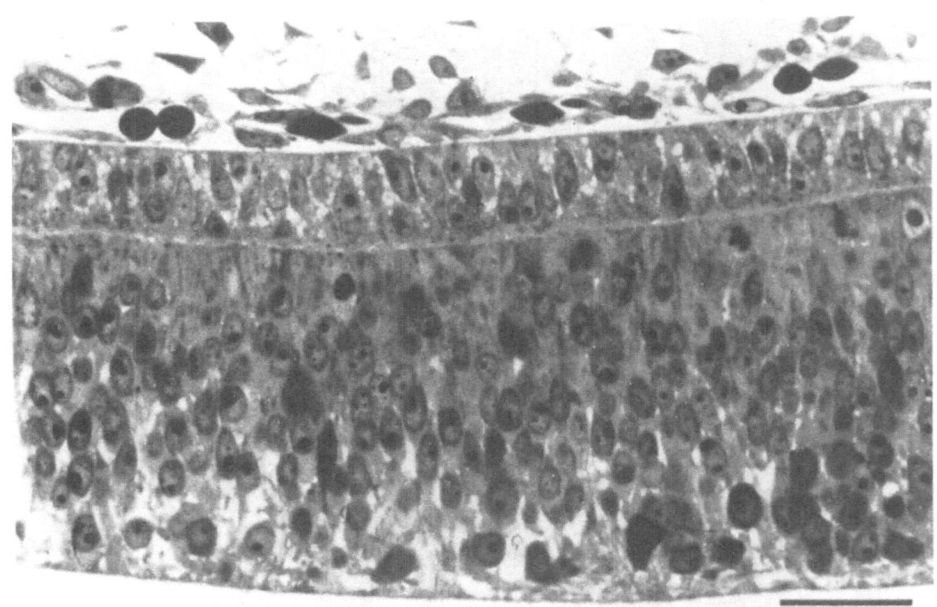

Fig. 7. Semithin section of the central area, of the same eye as in Fig. 6, at a higher magnification. Wide intercellular spaces can be seen near the inner limiting membrane. Young unipolar ganglion cells can be recognized from their circular appearance. *Bar* 30 μm

stratum gradually approaches the future outer stratum. At stage 15 (after 55 hours of incubation, Fig. 4) the inner stratum (retina) partially contacts the outer stratum (pigment epithelium). The lens epithelium forms a pit which is open towards the body surface. At stage 18 (only 15 hours later) the edges of the lens have come together and fused. The lens now forms a vesicle which is covered by the ectoderm (Fig. 5). The space between lens and retina is increased. The optic ventricle has disappeared except for small regions at the optic cup's margin. Retina and pigment epithelium are concave laterally along their whole extent; the radius of curvature decreases from the centre to the periphery. This situation is changed at stage 20 (Fig. 6); in the central region, retina and pigment epithelium temporarily have a negative radius of curvature. In this region the formation of the choroid is most advanced; many mitoses can be seen in the central retina while they decrease rapidly towards the periphery.

No mitoses can be found at the retinal margin in this section (Fig. 6). It is in the central region that the first ganglion cells can be detected near the inner limiting membrane. Ganglion cells have a spheroidal appearance in contrast to the densely packed bipolar cells in the ventricular zone (Fig. 7). They are surrounded by large intercellular spaces.

The ultrastructure of ganglion cells seems to be relatively undifferentiated at these early stages of development (Fig. 8). A thin cytoplasmic rim surrounds the spheroid nucleus. The cytoplasm is densely packed with ribosomes, but it contains only a few mitochondria and ergastoplasmic profiles. The basal processes of these young neurones are attached to the inner limiting membrane (Figs. 8,9). These attachments can be maintained even during the early phase of axonal outgrowth. This has been seen not

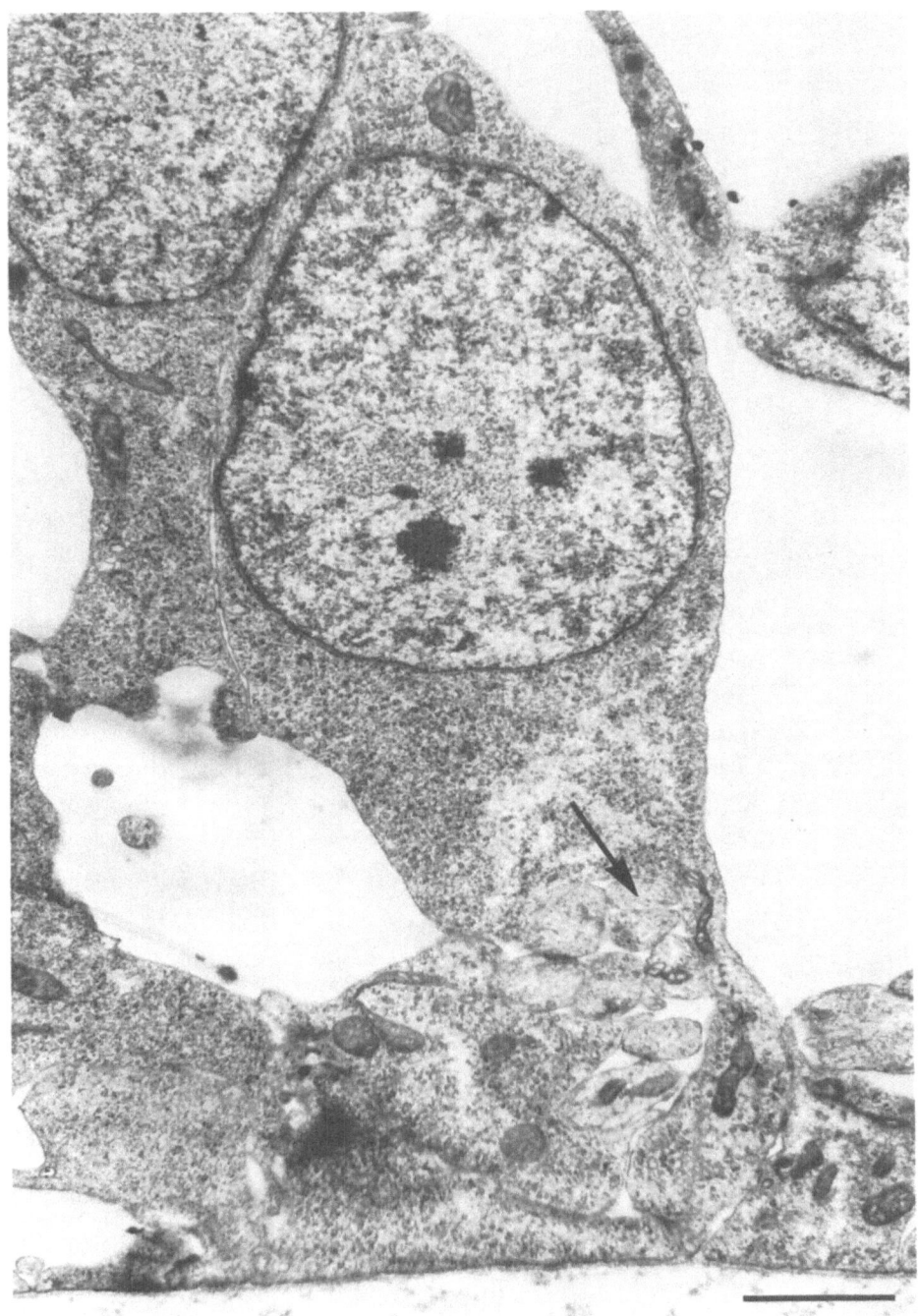

Fig. 8. A ganglion cell in the central retinal area at stage 20. The cytoplasm, densely packed with ribosomes, contains only a few organelles. The basal process is still present. Next to this process a small bundle of axons can be seen (*arrow*). *Bar* 15 µm

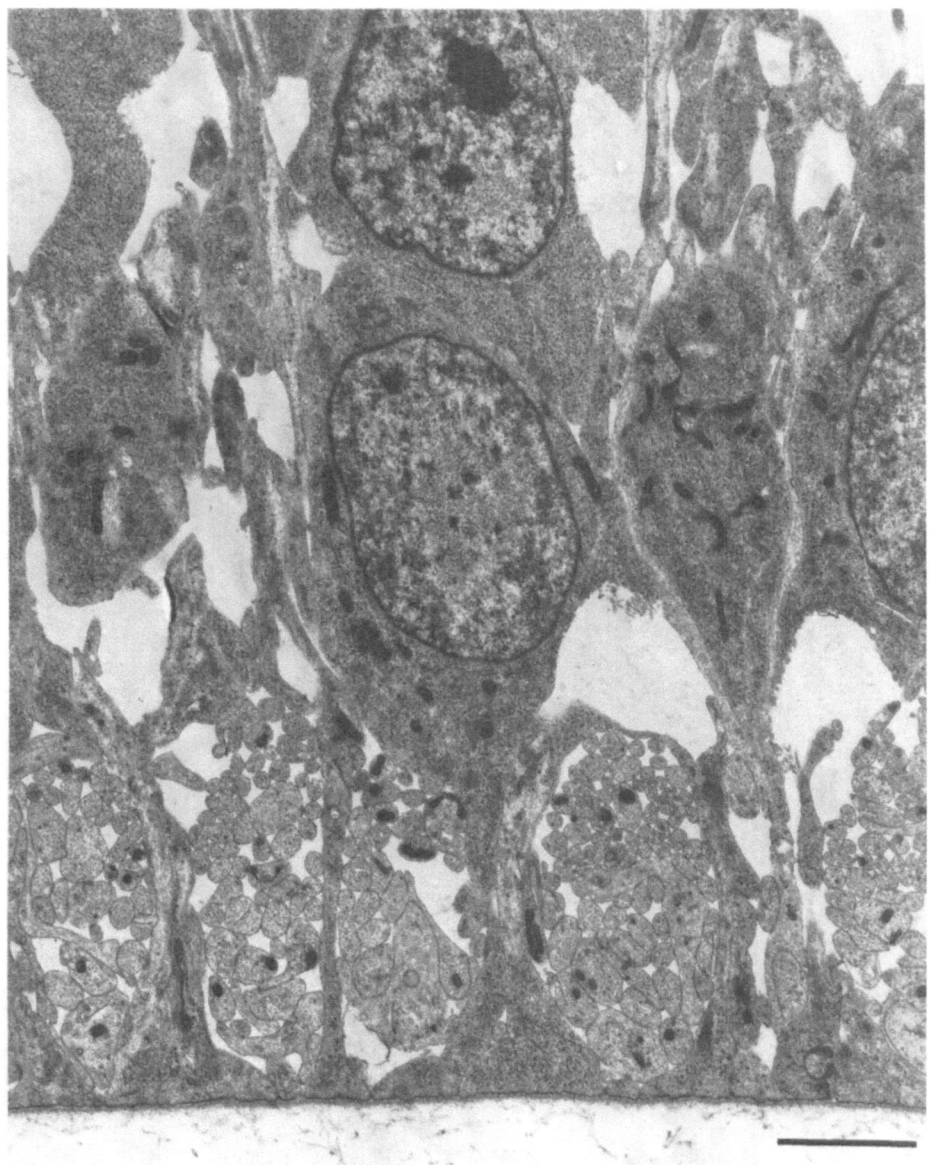

Fig. 9. Unipolar ganglion cells in the central retinal area on incubation day 5 (stage 26). Cell organelles are now concentrated mainly near the axon hillock. The fibre bundles can be subdivided into an apical portion in which thin axons predominate and into a basal portion in which growth cones predominate. *Bar* 2.5 μm

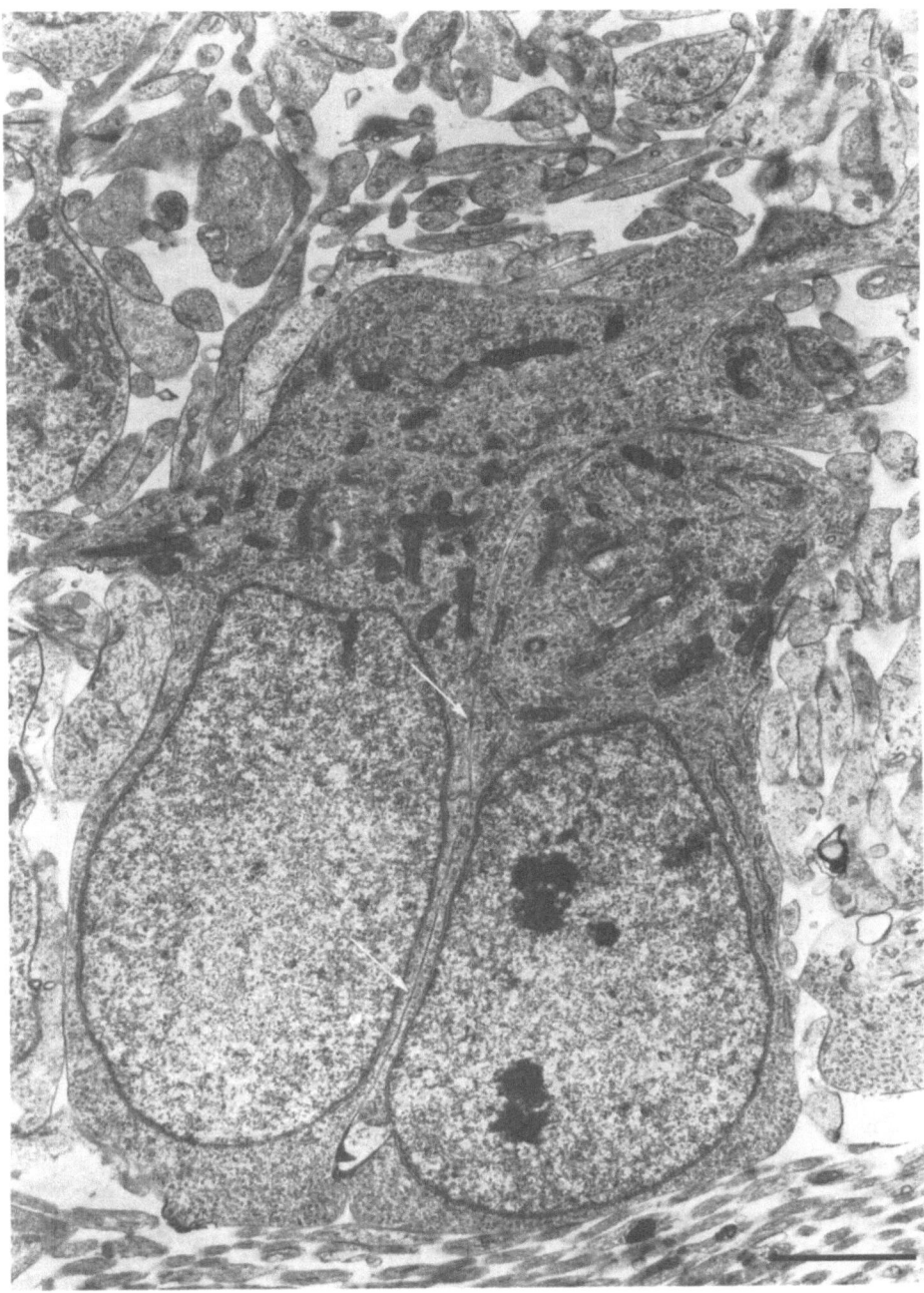

Fig. 10. A ganglion cell pair in the central retina on incubation day 8. *Arrows* indicate locations where no intercellular space seems to be present. Cell organelles are now concentrated in the apical part of the cell indicating that the cells are about to be transformed into their mature state. The dendrite has begun to sprout. *Bar* 2 μm

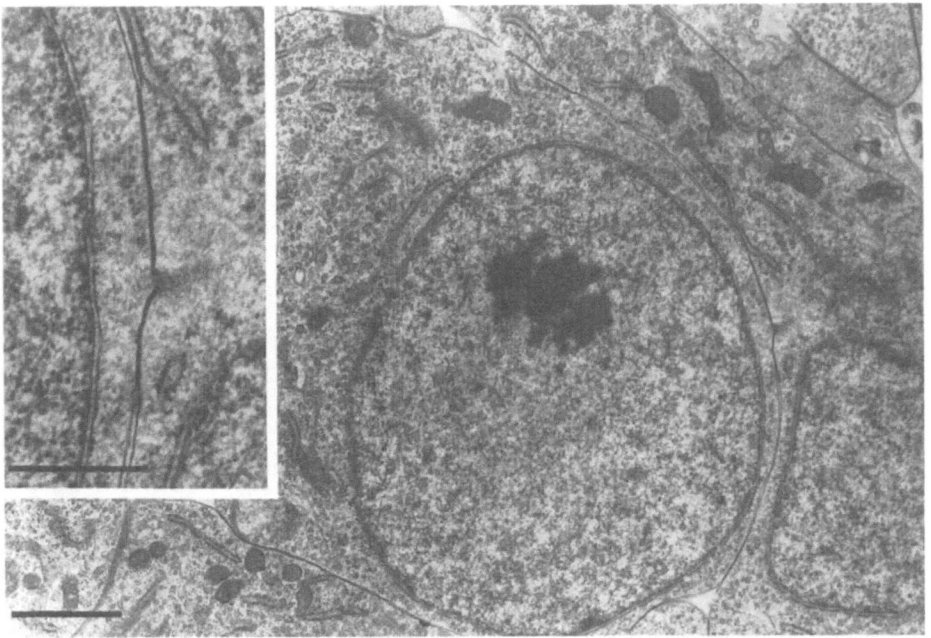

Fig. 11. A ganglion cell pair on incubation day 8. The plasma membrane is partially fused (*inset*). *Bars* 0.5 and 1.0 μm

only in our Golgi preparations but also in reconstructions from ultrathin serial sections in the mouse retina (Hinds and Hinds 1974).

The first cells that later become ganglion cells leave the mitotic cycle at stage 14 (after approximately 50 hours of incubation) according to ^{3}H-thymidine injections (Kahn 1974). With the electron microscope ganglion cell axons could be detected as early as stage 18. The axons immediately grow towards the optic fissure. They already form fascicles at this stage (Goldberg and Coulombre 1972). At later stages, fibres become progressively more orderly (Sect. 5). On incubation day 5 (stage 26), axons have already formed large bundles in which growth cones are located predominantly near the inner limiting membrane (Fig. 9). Mitochondria and ergastoplasm are now concentrated mainly near the axon hillock producing the typical appearance of a young neurone in its unipolar stage. On incubation day 8 more cell organelles are seen and they are now located in the apical part of the perikaryon where the dendrite has begun to sprout. A large Golgi apparatus can be seen in the same apical region near the nucleus (Fig. 10). Dendritic processes seem to grow in various directions; their range ist still small. These ultrastructural criteria characterize the multipolar stage.

Ganglion cells can often be seen as pairs at this stage of development. Cell membranes lie close together over the whole length of the perikaryon; the other side of each cell is loosely covered by glia (Fig. 10). At a higher magnification, we can see that the membranes are tightly attached at certain sites; occasionally they even fuse showing subjunctional specializations in the underlying cytoplasm (Fig. 11). In more mature stages glial processes seem to penetrate the space between these cell pairs (Fig. 12). Thus, individual ganglion cells largely become isolated from one another.

16

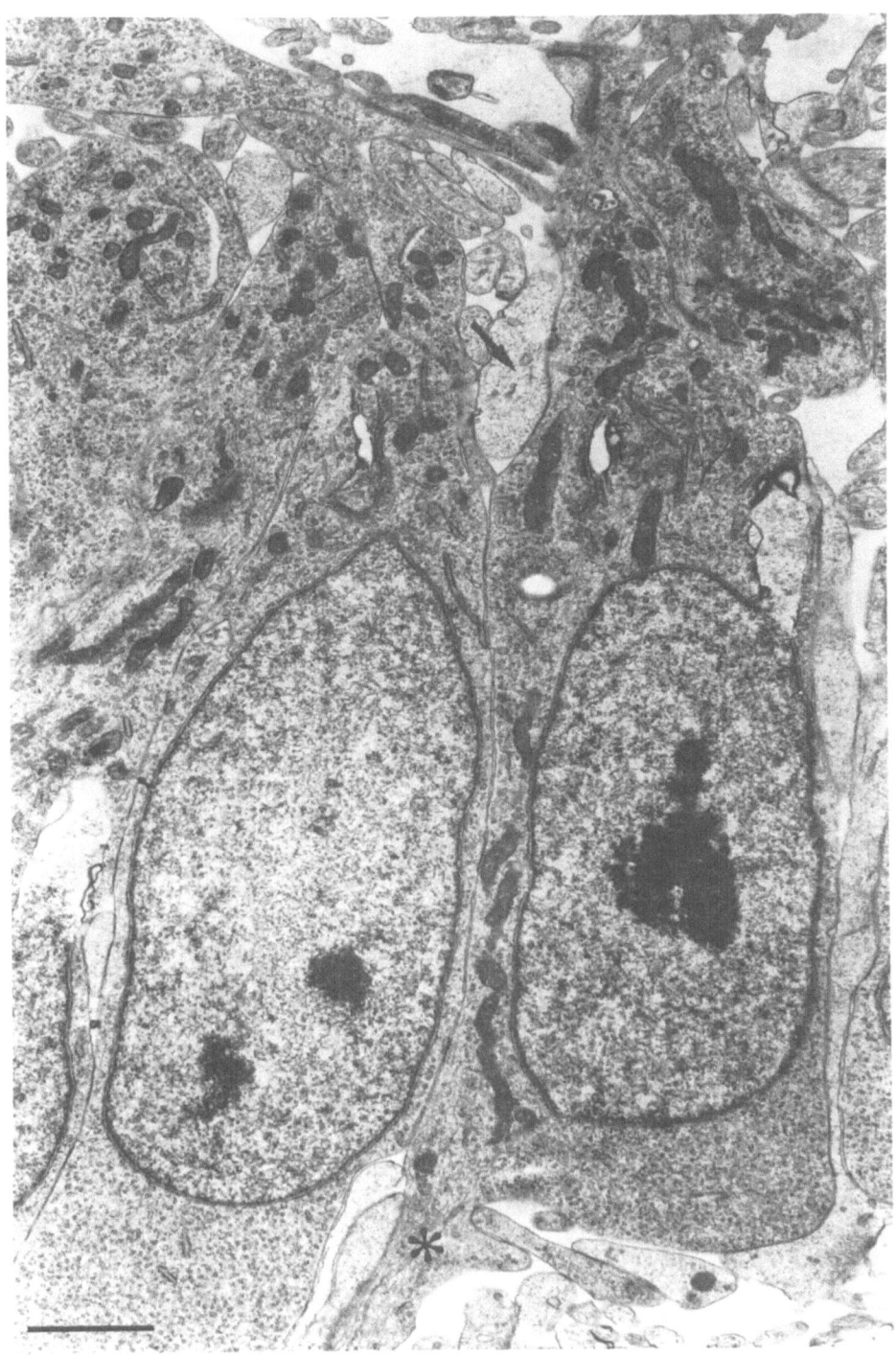

Fig. 12. Three tightly packed ganglion cells on incubation day 8. Glial cell processes seem to penetrate the intercellular space and to separate cells from each other (*arrow*). The *asterisk* identifies the axon of the right-hand cell. *Bar* 1.5 μm

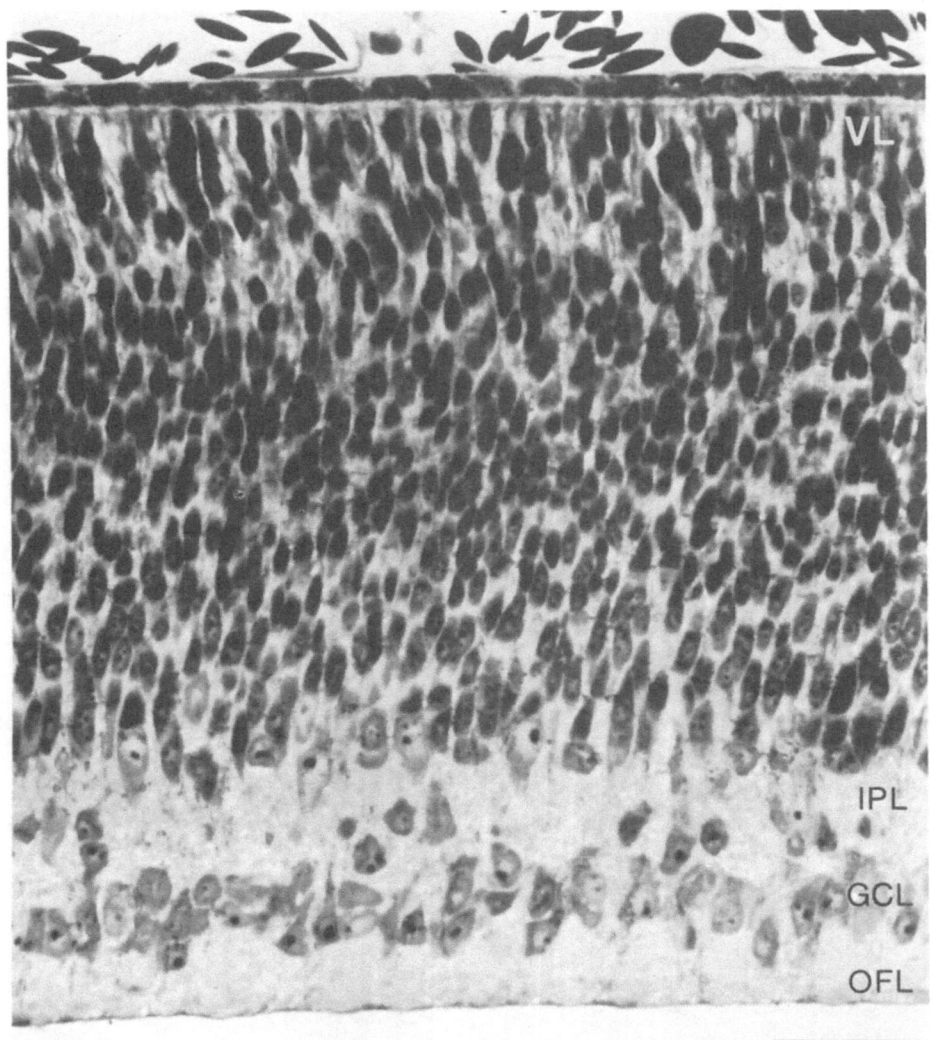

Fig. 13. Central retina on incubation day 8. The optic fibre layer (*OFL*), the ganglion cell layer (*GCL*) and the inner plexiform layer (*IPL*) can be recognized; the ventricular layer (*VL*) still mainly consists of immature, spindle-shaped cells. In the IPL a few cells appear to be migrating toward the GCL. *Bar* 25 μm

The generation and maturation of ganglion cells spreads from the central region to the periphery with time. The process of maturation, which is reflected in the size as well as structure of the inner plexiform layer, also proceeds from the centre to the periphery. On incubation day 8 we find a well defined inner plexiform layer in the centre of the retina (Fig. 13), whereas in the periphery no trace of an inner plexiform layer can be seen (see Fig. 14 in which a special reduced-silver stain (Rager et al. 1979) was used). The inner plexiform layer is divided up by the tangential spread of ganglion cell dendrites and amacrine cell processes in the sublayers. This stratification can

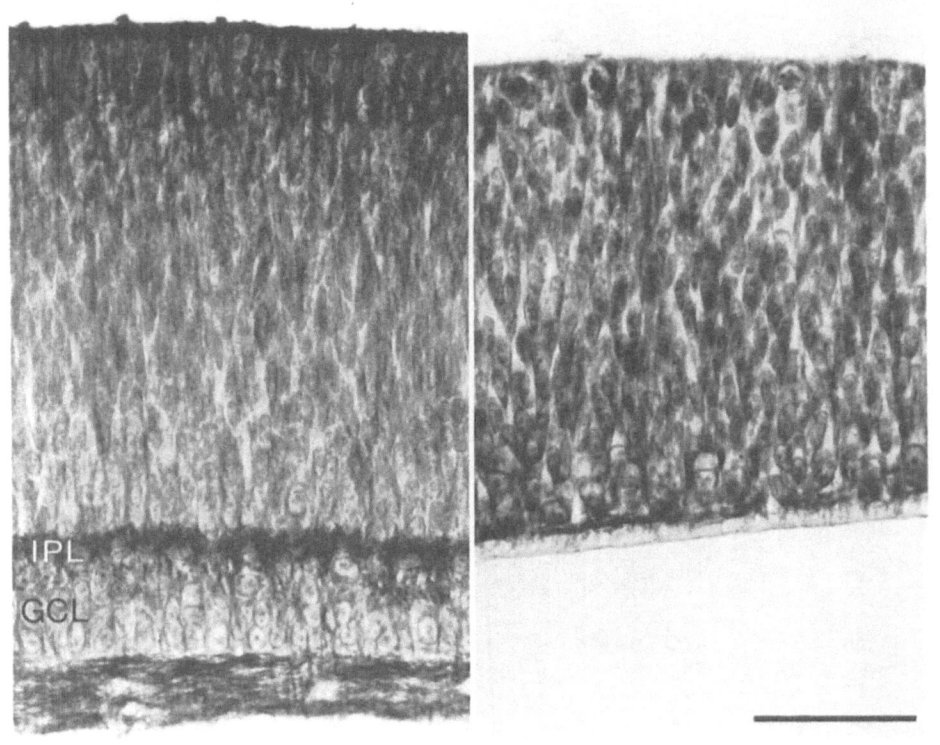

Fig. 14a, b. Central (a) and peripheral (b) retina on incubation day 8. Terminal branches are made prominent in the inner plexiform layer (*IPL*) with a special reduced silver stain (Rager et al. 1979). The IPL is already present in the centre of the retina, it is absent in the periphery. (*GCL*: ganglion cell layer) *Bar* 40 μm

already be recognized on incubation day 12 (dark tangential lines in Fig. 15). The thickness and stratification of the inner plexiform layer increase steadily until adulthood (Fig. 16). Uni- and multistratified cells were first described by Ramón y Cajal (1893), and more recently by Boycott and Dowling (1969).

3.2 Ganglion Cell Dendrites

When ganglion cells leave the mitotic cycle, they are initially bipolar in form with both an apical and a basal process. The apical process becomes progressively resorbed. Thus ganglion cells are transformed from bipolar to unipolar neurones. The pear-shaped perikaryon of the unipolar ganglion cell is limited by a smooth membrane. On incubation day 6, when axons of the earliest-generated ganglion cells arrive at the rostral pole of the optic tectum (De Long and Coulombre 1965; Goldberg 1974; Sect. 5), their perikarya enter a new stage of development. At this time short plump processes are formed which can originate from any site of the perikaryon and grow in all

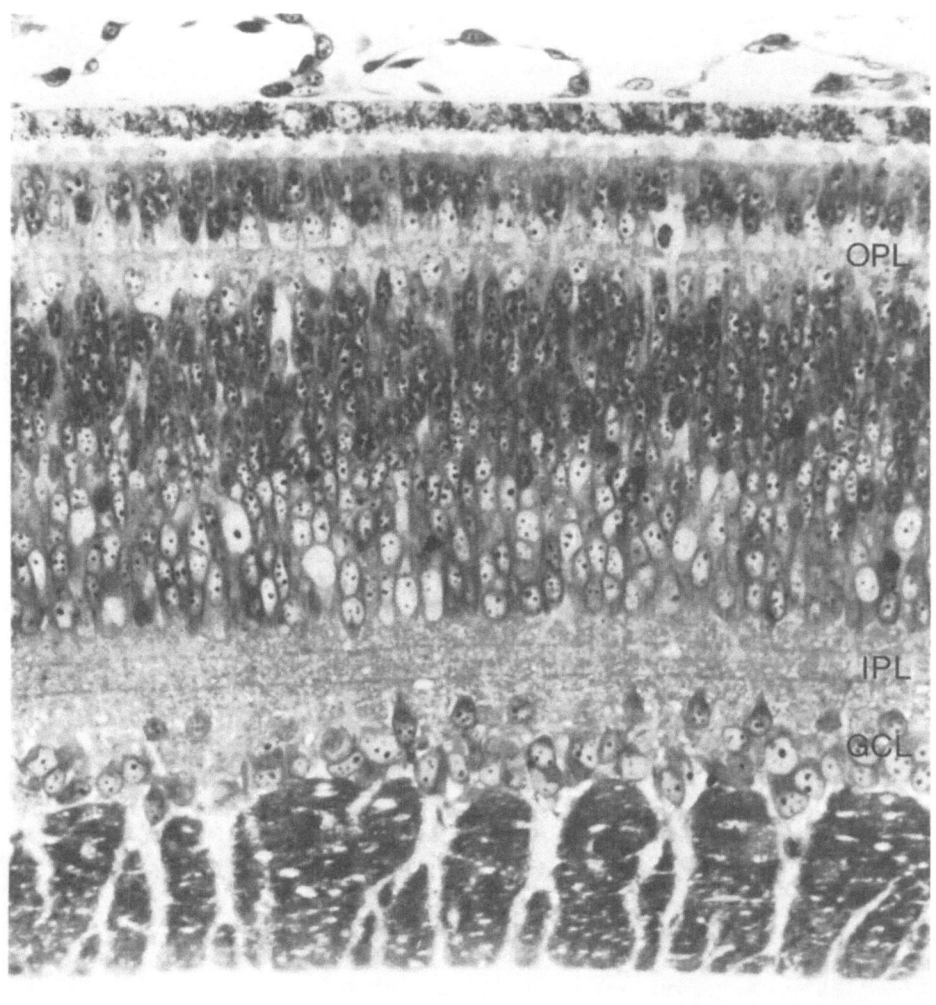

Fig. 15. Central retina on incubation day 12 (semithin section). Migrating cells are no longer present in the inner plexiform layer (*IPL*); all of them have arrived at the ganglion cell layer (*GCL*). Two tangential stripes can be seen in the IPL. They arrive from stratified ganglion and amacrine cell dendrites. The outer plexiform layer (*OPL*) and the inner segments of photoreceptors are just beginning to form. *Bar* 50 μm

Fig. 16. Central retina 22 days after hatching (semithin section). It has now attained its adult structure. Outer segments of photoreceptors invaginate the pigment epithelium (*PE*). In the outer plexiform layer (*OPL*) the pedicles of photoreceptors can be seen (*arrow*). The inner plexiform layer (*IPL*) has several tangential sublayers due to uni- or multistratified dendrites of ganglion and amacrine cells (*GCL*: ganglion cell layer; *OFL*: optic fibre layer). *Bar* 30 μm

PE

OPL

IPL

GCL

OFL

Fig. 16

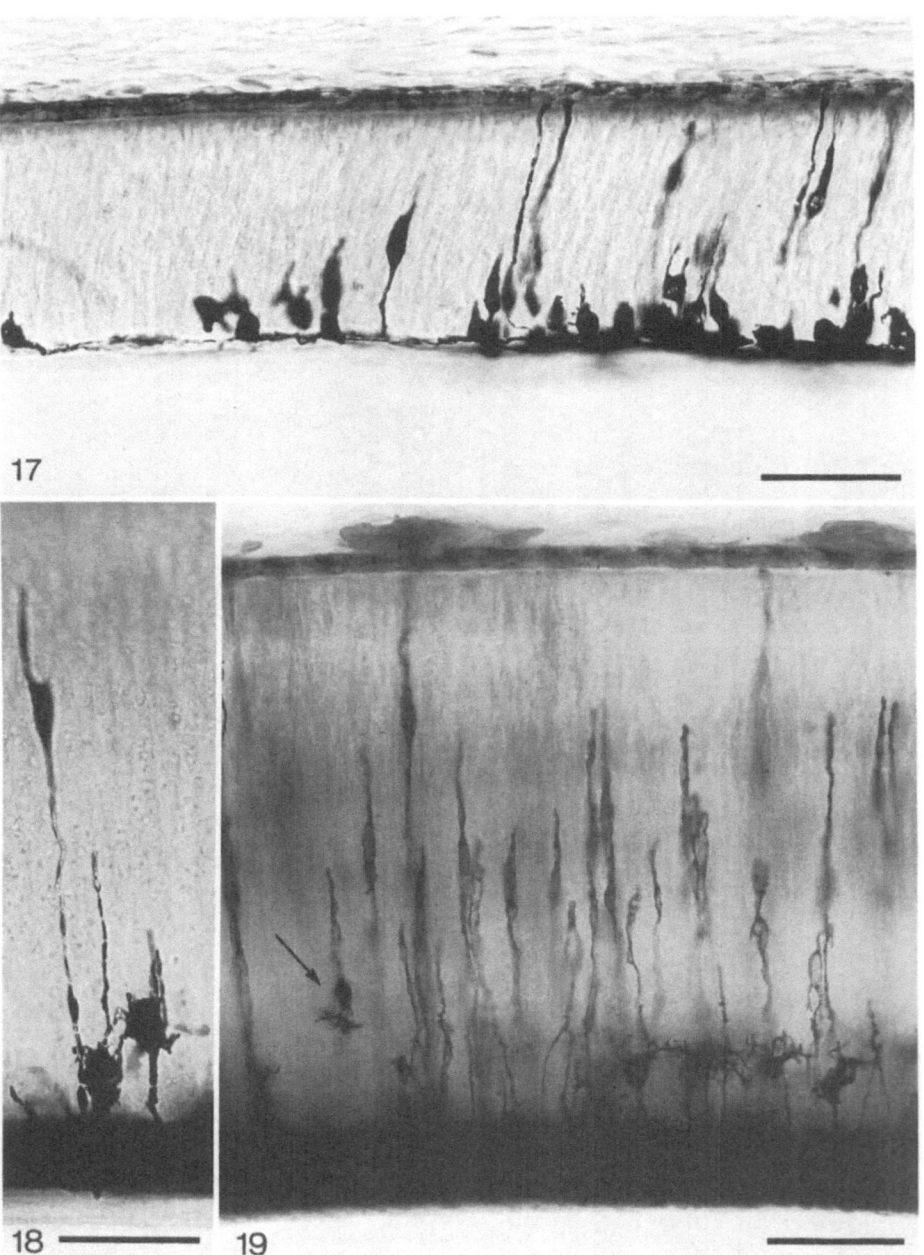

Fig. 17. Paracentral retina on incubation day 7 (Golgi-rapid method). Bipolar, unipolar and multipolar stages of ganglion cell development are present. *Bar* 50 μm

Fig. 18. Central retina on incubation day 6 (Golgi-rapid method). Ganglion cells in the multipolar stage having numerous short processes which can sprout from any portion of the perikaryon. *Bar* 40 μm

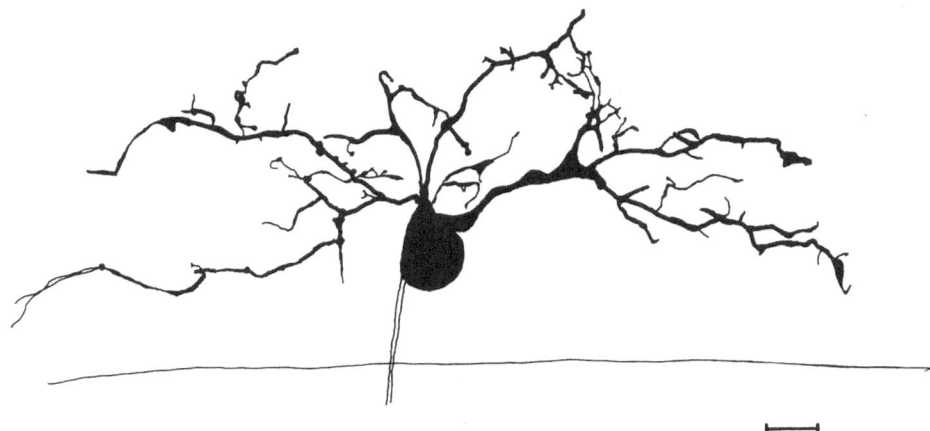

Fig. 20. A ganglion cell in the central retinal area on incubation day 9. Dendrites only arise from the apical pole of the cell, but a regular stratified pattern has not yet been reached. Some tips of dendrites have growth cones. Oil immersion, drawn with the aid of a camera lucida. *Bar* 10 μm

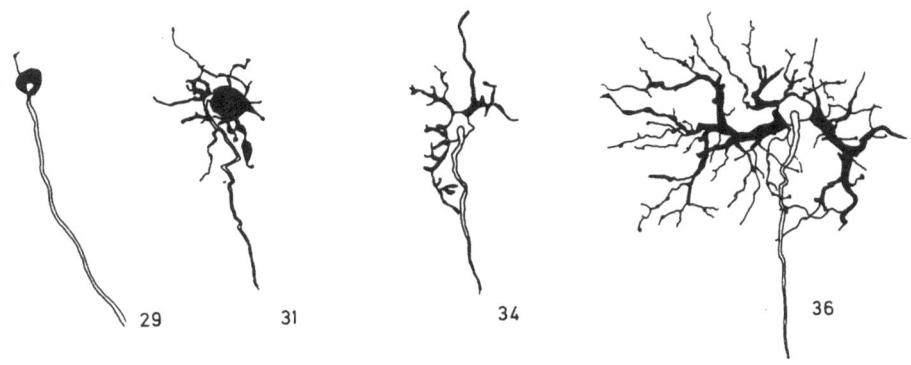

Fig. 21. Ganglion cells located in the central retina on incubation days 6 (stage 29), 7 (stage 31), 8 (stage 34), and 10 (stage 36). Golgi-impregnated whole mounts, oil immersion, drawn with the aid of a camera lucida. At stage 31, most of the processes are still plump and are in the same focal plane as the perikaryon. At stage 34, the final dendrite is about to be formed; it ramifies in a focal plane deeper than that of the perikaryon, so the dendrite is drawn in solid black and the perikaryon and axon are only outlined. *Bar* 10 μm

Fig. 19. Central retina on incubation day 8 (Golgi-rapid method). Ganglion cells are beginning to sprout their final dendrite. The maturation of amacrine cells (*arrow*) is practically synchronous with that of ganglion cells. *Bar* 50 μm

Fig. 22. Maturation of ganglion cells as a function of retinal excentricity seen in a Golgi-impregnated whole mount on incubation day 7 (stage 31). *Insets* give the retinal position (*small circle*) of the cells drawn above relative to the central *stippled* area and to the papilla. Oil immersion, *Bar* 10 µm

directions (Figs. 17,18). Also, fine spiny processes can often be seen. During incubation day 7, these processes increase in length and begin to branch. We define the period in which short ubiquitous processes can be seen as the "early multipolar stage." The electron microscopic criteria are given in Sect. 3.

The appearance of dendrites changes rapidly from incubation day 8 onwards (Fig. 19). Processes do not extend out in all directions any more; there is a single main dendrite at the apical pole of the cell. It first grows radially and then mainly tangentially producing the stratified appearance of the inner plexiform layer. The other processes of the early multipolar stage disappear. The transition from the multipolar to the mature form of the cell occurs very rapidly. Only one day later (incubation day 9) we see cells located in the central retina which show a single dendrite starting from the apical pole and then spreading tangentially (Fig. 20). These apical dendrites then continue to increase in size and ramification.

The description of the formation of ganglion cell dendrites obtained from radial sections is confirmed by Golgi-impregnated whole mounts. Four ganglion cells located in the central retinal area are drawn in Fig. 21. They are typical representatives of maturing cells on incubation days 6, 7, 8 and 10. The progress of maturation from

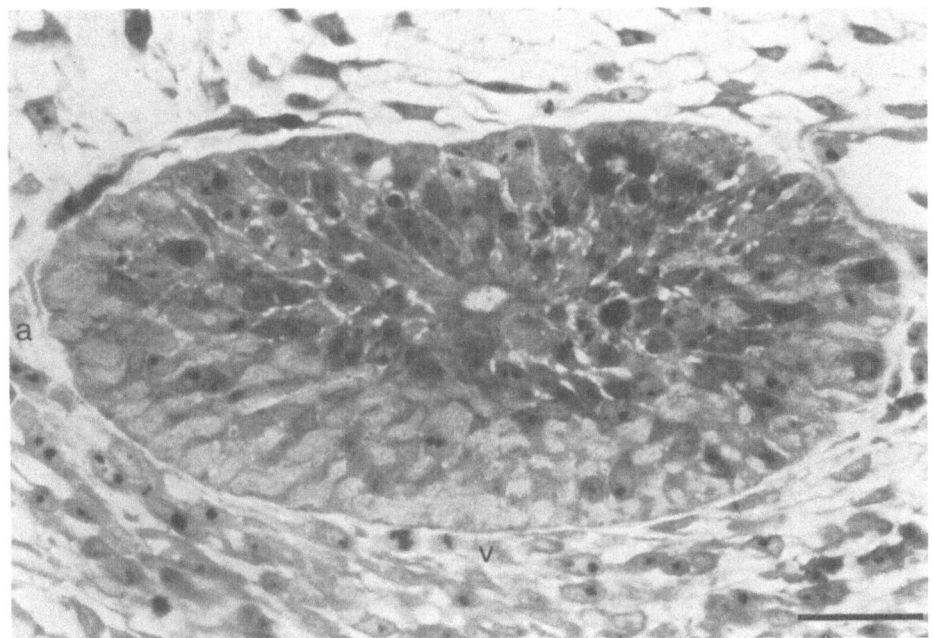

Fig. 23. Cross-section (semithin) of the optic nerve on incubation day 5. The lumen of the optic stalk is still open. The dorsal third is filled with undifferentiated neuroepithelial cells. The anterior (*a*), ventral (*v*) and posterior parts contain pale areas. These areas represent fibre bundles (see the electron micrographs in Fig. 26 and 27). Along the pial border the fibre bundles consist mainly of growth cones, while thin axons predominate in centrally located bundles. *Bar* 25 μm

central to peripheral retinal regions is illustrated in Fig. 22 where ganglion cells at progressively more eccentric positions in the same retina are drawn (incubation day 7).

Thus, the final maturation of ganglion cell dendrites seems to coincide with the arrival of fibres at their termination area and the onset of their branching process. This is shown and discussed below in Sects. 5 and 6.

3.3 Ganglion Cell Axons

On incubation day 3 the first ganglion cell axons enter the optic stalk which still has a tubular connection with the diencephalon. The number of newly formed ganglion cells doubles every 12 hours from incubation day 4 to incubation day 7 (Kahn 1973). Their axons fill the optic stalk rapidly. At the same time the fraction of undifferentiated neuroepithelial cells decreases. More than one-third of the cross-sectional area of the nerve is filled with these cells on incubation day 5 (Fig. 23). During the following days the area occupied by neuroepithelial cells is reduced drastically. Only a small rim remains in the dorsal part on incubation day 8 (Fig. 24) and it disappears one day later. Neuroepithelial cells are thinned out partly by invading fibre bundles and partly by degeneration (Fig. 25). The mechanisms responsible for this type of degeneration are not yet known.

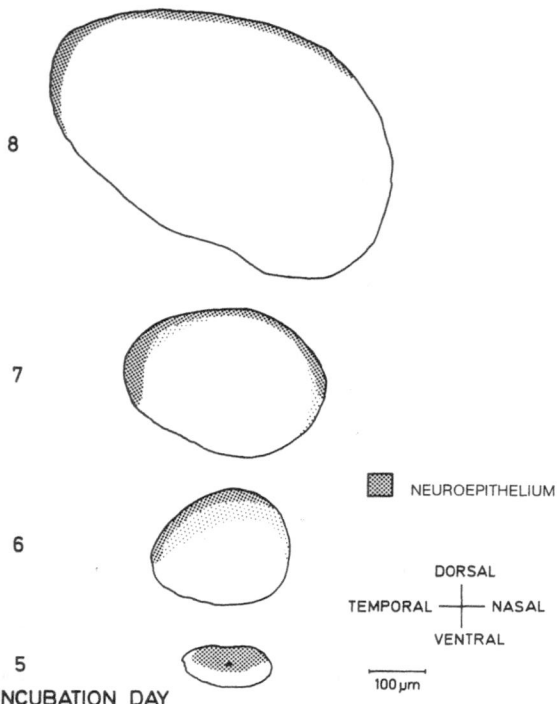

Fig. 24. Cross-sections of the optic nerve from incubation day 5 to 8 are drawn at the same magnification and orientation. The volume fraction occupied exclusively by neuroepithelial cells (*densely stippled area*) is gradually reduced. The transitional zone, where neuroepithelial cells are about to become interspersed with fibres, is *lightly stippled*

□ NEUROEPITHELIUM

DORSAL

TEMPORAL ——┼—— NASAL

VENTRAL

├——————┤
100 μm

8

7

6

5

INCUBATION DAY

During the early period of development, thin axons and growth cones can both be seen in the optic nerve. The cross-sections of growth cones are much larger than those of axons. Occasionally they have lateral processes in various directions and, therefore, have a multiform appearance (Figs. 27, 28). The packing density of microtubules is smaller in growth cones than in axons. Occasionally, we see large vesicles containing an electron-dense core (Fig. 28). We made photomontages composed of electron micrographs taken from optic nerve cross-sections of 4- and 5-day-old embryos at increasing distances from the retina. It is clear from such montages that centrally located fibre bundles consist mainly of thin axons (Fig. 26), whereas peripheral bundles contain mainly growth cones (Fig. 27). The significance of this phenomenon is discussed below in Sect. 5. However, there is no absolutely homogeneous population in central or peripheral regions. As we have already seen in the retina, younger cells grow between neurones which have reached their final position two to three days earlier (Fig. 13). Consequently, their growing tips reach any given position in the optic fibre pathway later and they are interspersed between homogeneously thin axons (Fig. 28). However, the percentage of such delayed growth cones is small; although thick profiles and growth cones are conspicuous in Fig. 28, they constitute only about 5% of the fibre population.

Growth cones represent a considerable fraction of fibres until incubation day 10. However, by incubation day 12 optic nerve fibres show a very homogeneous appearance (Rager 1976a, Figs. 4, 5). They are packed in large bundles which are separated by only a few glial processes. The diameters of 96% of fibres are less than 0.25 μm. A somewhat greater number of thicker axons appear on about incubation day 14. Then,

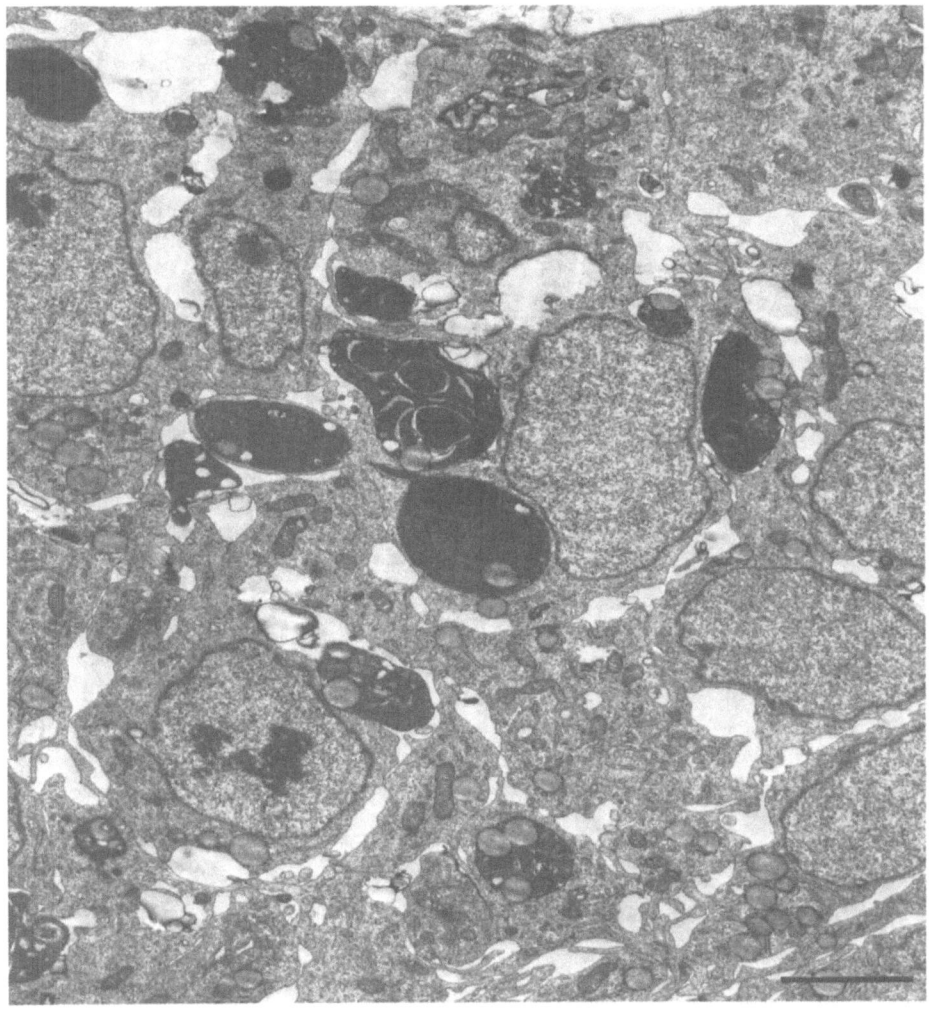

Fig. 25. Dorsal area of the optic nerve cross-section on incubation day 5. This area is exclusively occupied by neuroepithelial cells. A number of dark profiles can be seen both inside and outside cells. They resemble the debris of degenerating neurons. *Bar* 2.5 μm

as the oligodendroglia processes multiply, the bundles become finer and finer. On incubation day 15 these fine glial processes start to envelop the thicker axons. As the number of wrappings increases, the glial cell cytoplasm gradually disappears so that finally only double membranes and glial tongues can be seen. On incubation day 18, when photoreceptors can be activated by light stimuli for the first time (Rager 1979), only 1% of fibres are myelinated; around hatching this figure increases to 6%. Most of the myelination, however, takes place during the early post-hatching period. There is a considerable increase in fibre size (maximum diameter 3.5 μm) towards the end of the third month after hatching and nearly all fibres are then myelinated.

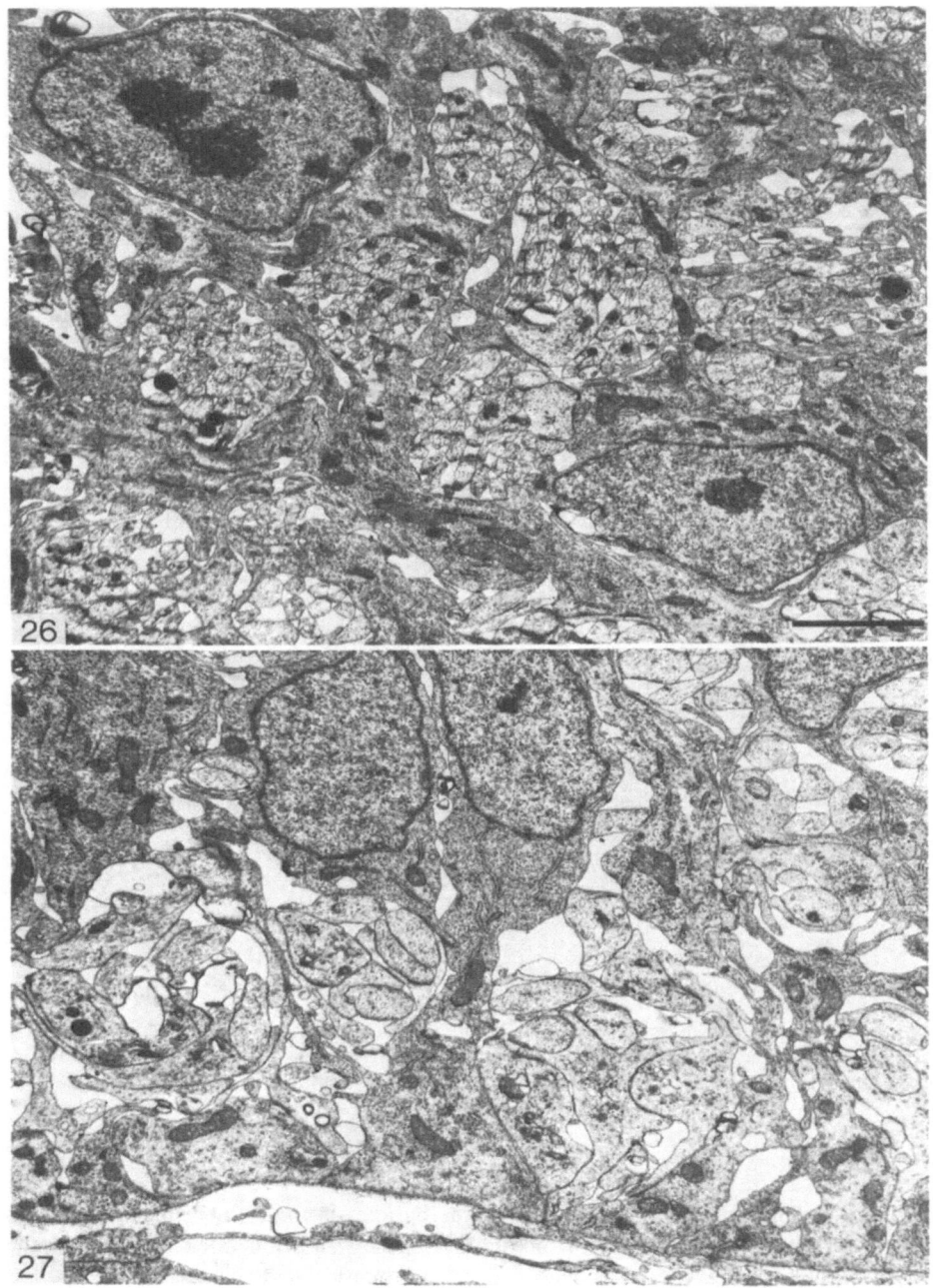

Fig. 26. In the central part of the fibre area of the optic nerve (incubation day 5) bundles of thin axons predominate. *Bar* 2.5 μm

Fig. 27. Small bundles consisting mainly of growth cones are found near the pial border. Same ultrathin section and magnification as in Fig. 26

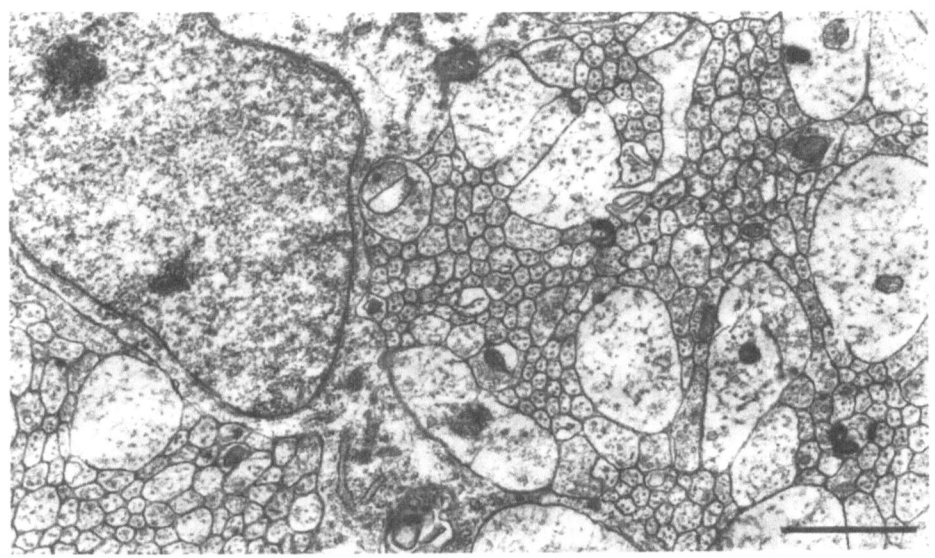

Fig. 28. Central core of the optic nerve on incubation day 7. A few thick profiles and growth cones can be seen amongst large bundles of thin axons. These are approximately 5% of the total. *Bar* 1.25 µm

Table 2. Development of myelination

Day of ontogenesis	Fibres counted No.	Myelinated fibres %
14	22 900	0
15	12 624	0.03
16	14 450	0.2
17	12 351	0.8
18	22 967	1.2
21	5 467	6.1
26	13 914	10.6
28	9 462	13.1
35	10 136	18.8
50	7 887	35.0
78	8 640	64.3
119	4 543	93.7

Table 2 and the histograms shown in Fig. 29 give a quantitative summary of this course of development. Fibres were classified according to their minimum diameter (class width 0.5 µm). The frequency of each diameter class is indicated on a logarithmic scale. A continuously increasing percentage of axons becomes thicker from incubation day 12 onwards when all fibres are small and unmyelinated and growth cones have disappeared. Myelinated fibres can be seen first on incubation day 15. However, at each stage, the distribution of the whole fibre population is unimodal.

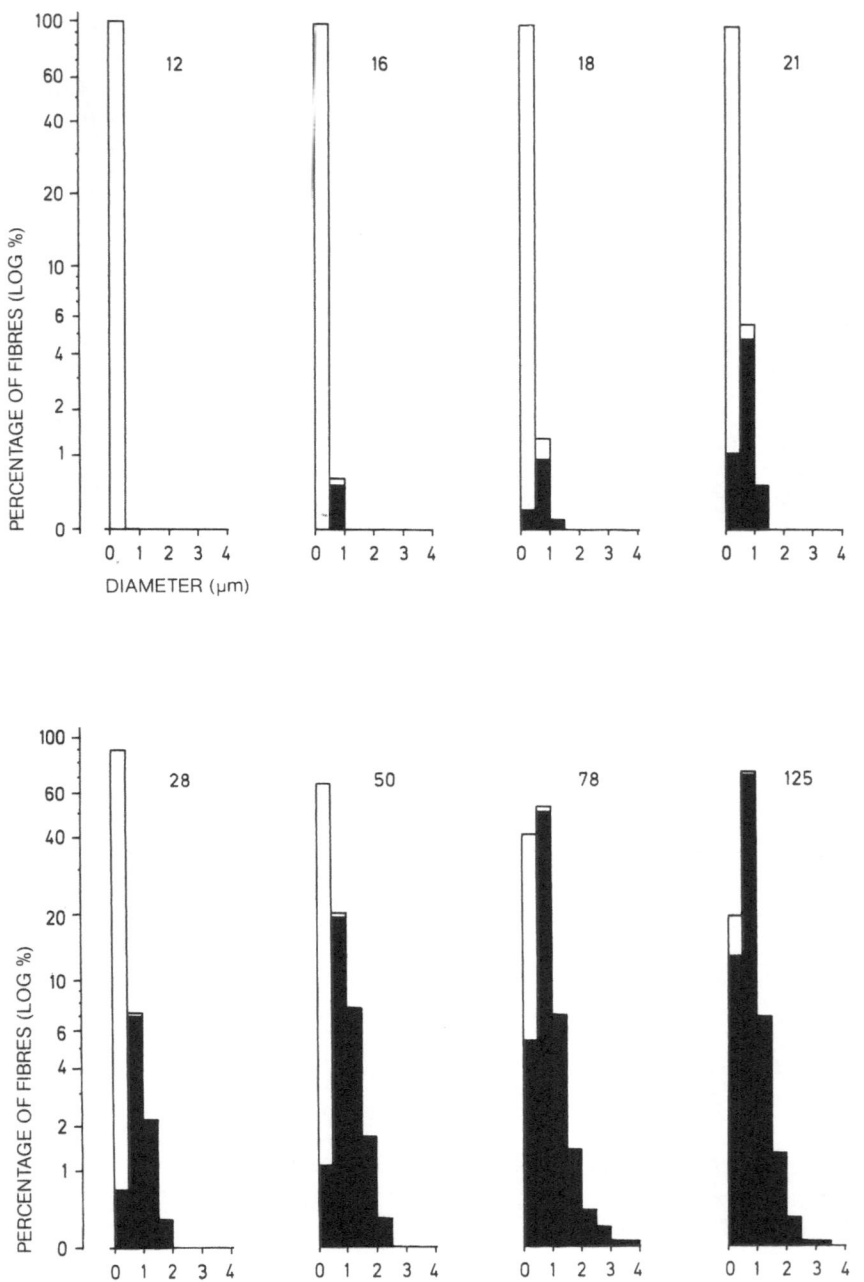

Fig. 29. The development of fibre diameters and of myelination. Diameter classes having a width of 0.5 μm are plotted as histograms on a logarithmic percentage scale. The age of the animals is given in days of ontogenesis. *White bars* represent unmyelinated and *black bars* myelinated fibres

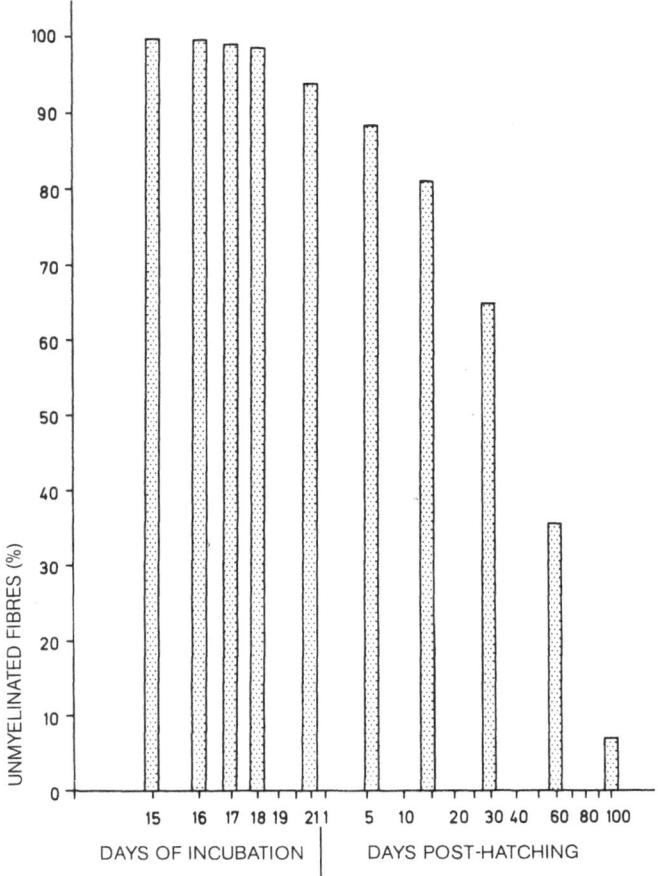

Fig. 30. The percentage of unmyelinated fibres is plotted against developmental age (days of onto-genesis); time scale is logarithmic

Three months after hatching the peak value lies at 0.9 μm with a standard deviation of 0.4 μm. The increase in the percentage of myelinated fibres during this period of development is shown in Fig. 30. At the end of the third month 94% of fibres are myelinated.

3.4 Synaptogenesis in the Optic Tectum

Axons of the earliest-generated ganglion cells arrive at the central tectal area on incubation day 8. They start to ramify and to invade the tectal cortical plate immediately (Sect. 5). There is some evidence from Golgi impregnations that retinal fibres can form end-knobs one day later and contact the dendritic surfaces of tectal cells (Rager and von Oeynhausen 1979). With the electron microscope quite a number of axon terminals can be seen in the upper tectal layer. These terminals contain many vesicles. In a few cases vesicles tend to aggregate (Fig. 31), but only very rarely did we find

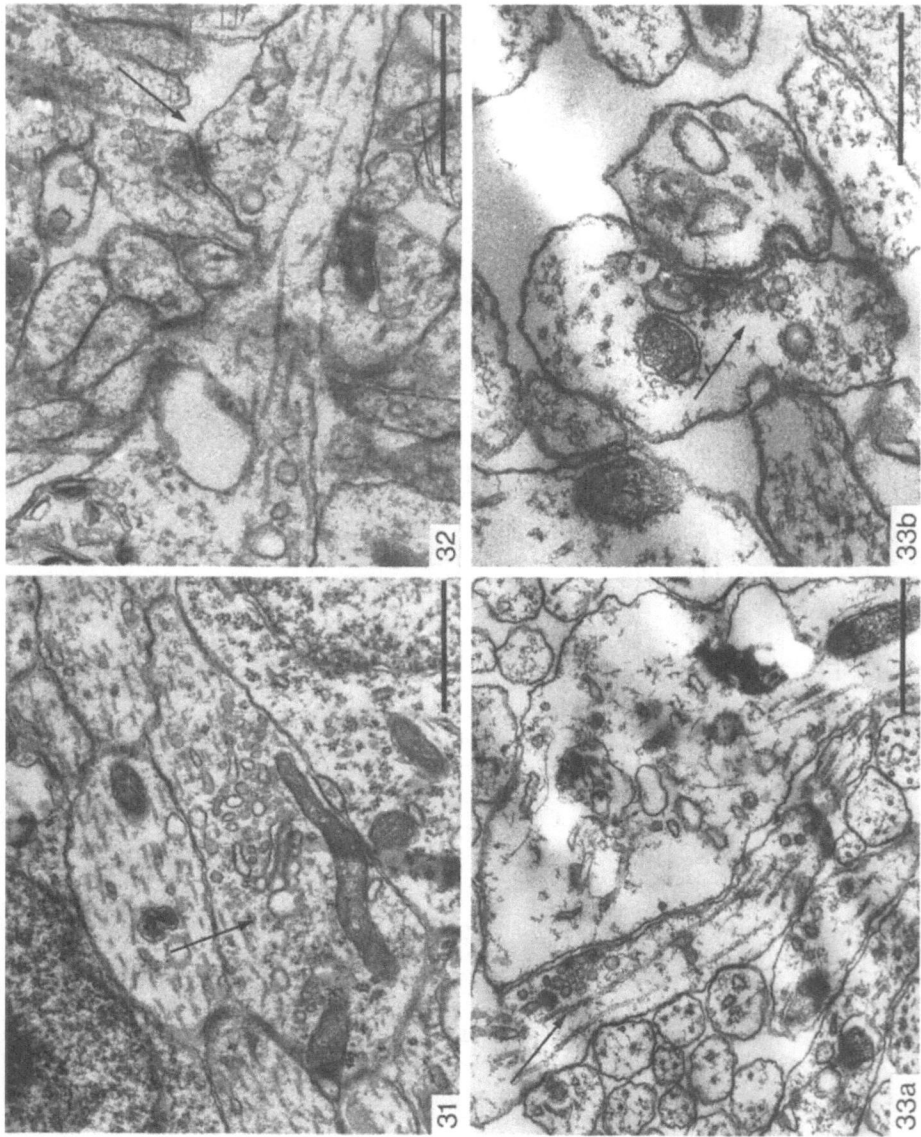

Fig. 31. Axon terminal in the upper cortical plate on incubation day 9 (*arrow*). It contains a number of vesicles. *Bar* 1 μm

Fig. 32. A tiny synapse (*arrow*) located in the upper cortical plate of the optic tectum on incubation day 9. The origin of its axon is unknown. *Bar* 0.8 μm

Fig. 33a, b. Synapses found in the central tectal area on incubation day 10 (*arrows*). They are located in the stratum opticum (a) and immediately below the stratum opticum (b). In both cases the synapses probably originate from retinal axons because of their location. *Bar* in (a) 0.6 μm; in (b) 0.4 μm

terminals having vesicles concentrated near specialized membranes (Fig. 32). On incubation day 10, synapses can be found in the inner half of the stratum opticum (SO) and immediately below it (Fig. 33). A recent quantitative electron microscopic analysis of all tectal layers on incubation day 10 (Rager, to be published) shows that the frequency of occurrence of synapses is highest in layers VIII and VII (designation according to La Vail and Cowan, 1971a). This is consistent with the ingrowth of fibres seen in our Golgi-impregnated material. From incubation day 11 onwards, synapses can be found fulfilling the following criteria supporting their identification as retinal synapses:

1. They were localized within the SO or at least in the dendritic field of the first post-synaptic neurons.

2. The pre-terminal axons could be traced for a distance long enough to discover their origin in the SO either by direct observation or indirectly from their orientation.

3. The contacts had all the morphological properties of a synapse: pre-synaptic vesicles, a synaptic cleft, and pre- and post-synaptic membrane specializations (Rager 1976b).

These results suggest that there is a continuous process; after their arrival retinal fibres invade the cellular tectal layers and progressively form synaptic contacts.

4 Physiogenesis

Previous electrophysiological investigations of the development of the visual system of the chicken have concentrated almost exclusively on the development of the electro-retinogram and of visually evoked responses (Peters et al. 1958; Garcia-Austt and Patetta-Queirolo 1961; Witkovsky 1963; Pisareva 1965; Blozovski and Blozovski 1968; Sedlacek 1969; Ansinn et al. 1969). These investigators and Rager (1979) showed that photoreceptors can only be excited by light stimuli from late on incubation day 17 onwards. Thus, electrical stimulation has to be used to investigate the functional maturation of the retinotectal system at earlier stages of development. Our study was guided by the following questions: (1) When do axons begin to transmit signals? (2) When do retinotectal synapses start to function? (3) How are structural and functional development related to each other?

4.1 Development of Fibre Activity

4.1.1 Evoked Potentials

The optic nerve head in the retina was stimulated and records were taken from the tectal surface with a silver ball electrode, the indifferent silver ball electrode being placed in the beak lumen. The resistances of the silver ball electrodes were both low enough to permit noise-free recording of signals of only a few microvolts. In the older embryos, a control set of optic tract potentials was also recorded with micropipettes.

The youngest embryos from which we recorded were eight days old (Fig. 34). At this age, only a few retinal fibres have arrived at the central tectal area. Thus, at this

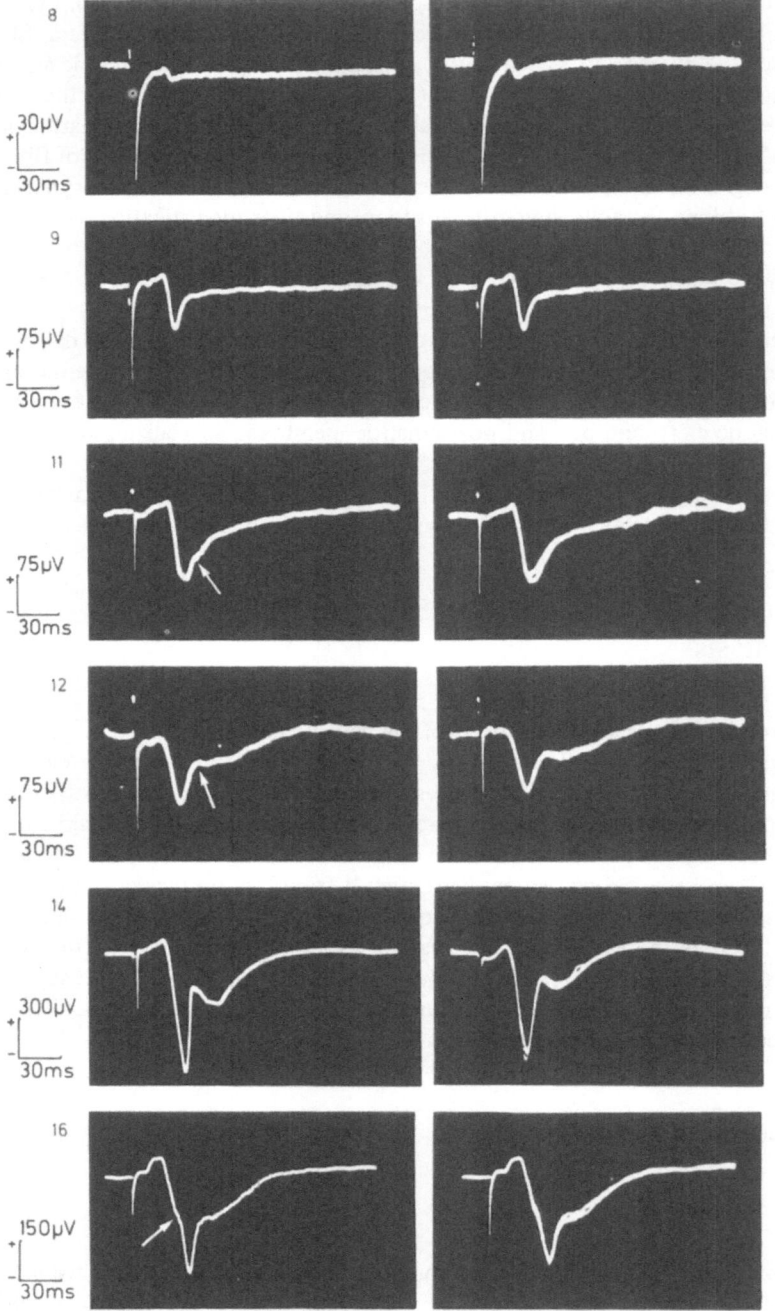

Fig. 34a, b. Evoked responses recorded from the surface of the central tectal area after electrical stimulation of the optic nerve papilla. a) shows single sweeps, b) shows groups of three sweeps superimposed. On incubation days 11 and 12 additional potentials are superposed on the repolarization phase (*arrows*). On incubation day 16 there is a kink in the depolarization phase (*arrow*); at this stage two fibre groups can be distinguished because of differences in conduction velocity

stage evoked responses can only be found in the anterior part of the tectum. Recording from the central tectal area only 24 h later, we observe a drastic change. The amplitude is greater by a factor of ten, even though the latency remains about the same. At later stages we see additional changes in potential (Fig. 34, arrows on curves recorded on incubation days 11 and 12) which are apparently due to post-synaptic potentials as discussed in Sect. 4.2.

At the onset of myelination, on incubation days 15 and 16, two fibre groups can be identified with differing conduction velocities (Fig. 34, incubation day 16, *arrow*). Further negative troughs appeared at later stages (near hatching time, see Rager 1976a). During post-natal development, the changes are mainly quantitative, i.e. as the myelination increases the latencies become shorter, the fibre action potentials become larger in amplitude. This was shown elsewhere (Rager 1976b) by comparing responses to ortho- and antidromic stimulation.

Younger embryos require a much greater stimulus strength than older ones to get the maximal response. In addition, the interstimulus interval had to be greatly prolonged (up to 20 s). As we have seen in Sect. 3, the large bundles of unmyelinated optic nerve fibres are only separated by few thin glial septa before the onset of myelination. Assuming that all fibres in such a bundle are activated on supramaximal stimulation, two possible explanations for the long refractory period would be:

1. Potassium accumulates in the extracellular space because of the presence of only a small number of glial processes and this results in a long-lasting diminution of the membrane potential (Orkand et al. 1966; Kuffler and Nicholls 1966).

2. Although action potentials can be generated at this stage, the ionic mechanisms intrinsic to the axon might not yet be fully developed.

At later stages fibre bundles become more and more split up by growing glial processes, and the appropriate interstimulus intervals become shorter and shorter. On incubation day 16, unmyelinated fibres already show properties similar to those of adult sC-fibres. We applied paired stimuli with varying intervals to the tectal surface and recorded the antidromic fibre potential at the optic nerve head. Fibres were able to follow the stimuli with interstimulus intervals as small as 20 ms without reduction in amplitude. In addition, supernormality, as described by Erlanger and Gasser (1937), was seen at interstimulus intervals between 100 and 40 ms (Rager 1976a).

4.1.2 Fibre Diameter and Conduction Velocity

The increase of fibre diameters with age has been shown in Fig. 29 where the whole fibre population was considered. However, when comparing conduction velocity with fibre diameter, it was necessary to limit the study to the fastest conducting fibres since they are the only ones whose conduction velocity can be measured with a sufficient degree of precision. In general, conduction velocity increases with fibre diameter. Thus, the conduction velocity of the fastest fibres was compared with the mean diameter of the thickest fibre class, the class width being 0.25 μm. Fibre diameters were measured from electron micrographs; they were not corrected for shrinkage or compression.

The observations made before incubation day 10 were not included, because there were too many growth cones. On incubation day 10, the overwhelming majority of

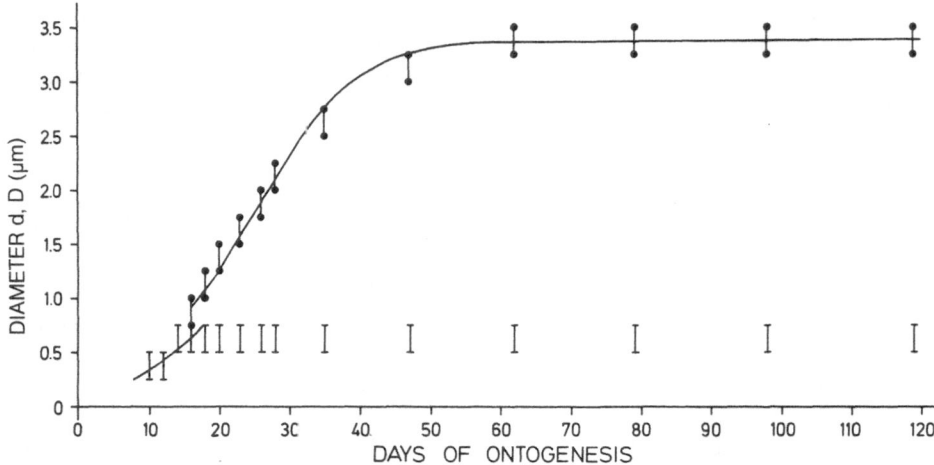

Fig. 35. Development of the largest diameter class as a function of time. The symbol ⊥̄ represents the largest myelinated fibres (D), the symbol ⊥ the largest unmyelinated fibres (d) from our measurements. The graph is obtained from the mathematical model outlined in 6.2. The discontinuity in this graph (incubation days 16 and 18) is due to the onset of myelination

fibres (97%) fall into the class of smallest diameters (0.0–0.25 μm, class 1), only 3% fall into class 2 (0.25–0.5 μm). On incubation days 12 and 14 the proportion of fibres in class 2 has increased to 4% and 9%, respectively. On days 14 and 16, a few fibres reach class 3 (0.5–0.75 μm). Unmyelinated fibres with larger diameters were never found, even three months after hatching. The proportion of unmyelinated fibres with diameters in class 3 reaches 1% at hatching time. It remains at this level even though the percentage of unmyelinated fibres in the whole population falls to a final value of 6%. This suggests that there is a critical diameter between 0.5 and 0.75 μm at which a fibre becomes myelinated.

Myelinated fibres are present in classes 3 and 4 (0.75–1.0 μm) on incubation day 16. Their diameters then increase, initially rapidly and later more slowly, and approach maximal size during the first half of the second month. The graph shown in Fig. 35 illustrates the changes of large fibre diameters with time which were predicted with the mathematical model derived in Sect. 6 and fitted to the measured values. For myelinated fibres the ratio $\rho = d/D$ of axon diameter d to fibre diameter D has an average value of 0.7 (Rager 1976a). Thus, the graph shows a discontinuity at the time when myelinated fibres first appear, since the diameter of myelinated fibres is, on an average, bigger than that of axons by a factor $1/\rho = 1.43$.

Conduction velocity is plotted against developmental age on a double logarithmic scale in Fig. 36. The velocity starts at 0.26 m/s on incubation day 8 and increases very slowly at first. As soon as myelination begins, the velocity increases rapidly. It eventually approaches a maximum value of 12–13 m/s. For unmyelinated fibres a curve based on a linear relation between velocity θ and fibre diameter d, i.e., θ proportional to d fitted the experimental values closely. For myelinated fibres, however, the experimental data reveal a quadratic relationship between velocity and diameter, i.e., θ varies as D^2. The values of diameter d and D are taken from Eq. 15 and 16 (Sect. 6) of the fibre-diameter function.

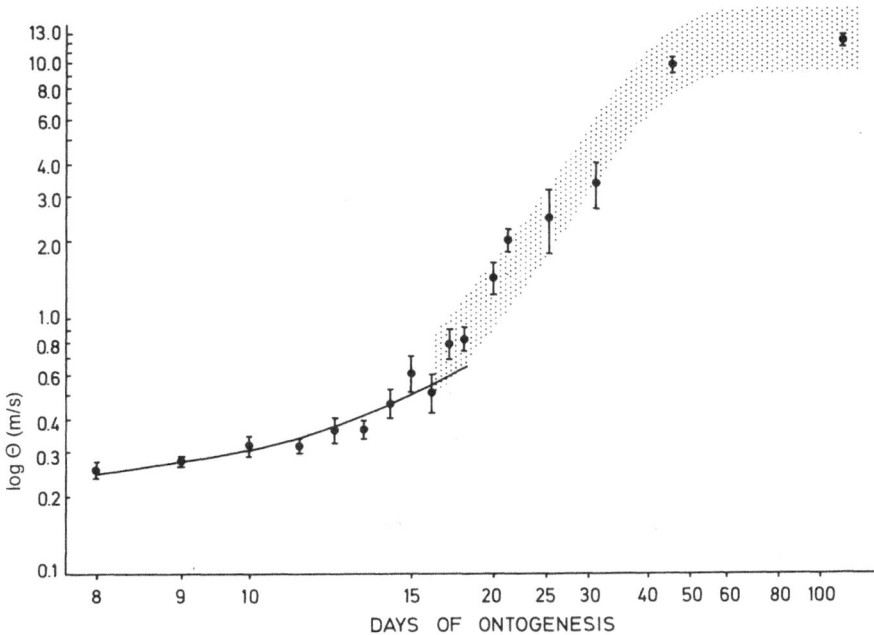

Fig. 36. The conduction velocity θ of the largest fibres plotted against time t on a double logarithmic scale. The measured values of unmyelinated fibres could be fitted best by a function assuming a linear relation between θ and fibre diameter d. For myelinated fibres a quadratic relation between θ and the outer diameter D gave the best fit. The conduction velocity would vary within a band indicated by the stippled area if ρ (the ratio of axon d and outer fibre diameter D of myelinated fibres) varied between 0.6 and 0.8

4.2 Physiology of Synaptogenesis

The evoked potential recorded from the optic tectum after electrical stimulation of the optic nerve is initially negative. By incubation day 11 the repolarization phase is interrupted and changes into a slow return to the base line (Fig. 34). These changes in the repolarization phase are much more pronounced on incubation day 12. The interrupted ascending phase is immediately followed by a new small, negative deflexion which slowly returns to zero. Two days later (incubation day 14) this second negative deflexion is more prominent. Thus, the compound fibre action potential is followed by additional potential changes. These additional potential changes can be distinguished even from the slowest fibre action potentials by the magnitude of their latencies (Sect. 4.1). The question now arises whether these additional potentials are post-synaptic in origin. This problem has been approached by applying the following techniques: (1) paired stimuli with varying interstimulus intervals, (2) combined ortho- and antidromic stimulation, (3) depth profiles of current sources and sinks, (4) single cell recordings. The results obtained with all of these techniques indicate that retinal fibres stimulated electrically are able to activate tectal neurones at least from incubation day 11 onwards (Rager 1976b).

   In a recent series of experiments (Rager and Kreische 1978; Rager et al., to be published) current-source-density analysis was used in the way proposed by Mitzdorf and Singer (1977). The central area of the optic tectum was penetrated radially with micropipettes. Evoked potentials were recorded after optic nerve stimulation at depth intervals of 25 μm. The second derivative of the potential with respect to tectal depth was used to estimate the local extracellular current density. This technique is applicable here because the avian optic tectum is a clearly laminated cortical structure and widespread synchronized activity then leads to a practically one-dimensional current flow. Below the well-defined afferent zone (SO) there are a number of tangentially oriented laminae (Ramón y Cajal 1911). However, most of the cells in these laminae and their dendrites are orientated radially (Sect. 5) and they thereby transmit the retinal input towards the tectal centre (Stone and Freeman 1971). As a consequence of this structural arrangement, excitation of a large number of synapses located on dendrites of radial tectal cells should momentarily set up a dipole. A current sink generated in the superficial layers by the dendritic excitatory post-synaptic potentials would draw current from the corresponding perikarya (Chung et al. 1974). These perikarya are located in deeper layers and should give rise to a deep zone of current sources. This has been observed in the tectum of the adult pigeon (Holden, 1968a, b; Stone and Freeman 1971). Thus, if such a potential reversal occurs at one particular stage of development, it would suggest that synaptic transmission is effective from then on.

In the chick embryo a potential reversal could be found even on incubation day 9 which was the earliest stage at which such an analysis was done. At this stage fibres do invade the cortical plate (Sect. 5), but synapses can be found only on very rare occasions. We are not sure whether these synapses originate from retinal fibres (Sect. 3). Thus, a synaptic activation due to retinal input is very unlikely at this early stage of development. How else can this phenomenon be explained? The Golgi-impregnated material (Sect. 5) reveals that growth cones are present not only in the SO, but also in deeper tectal layers. They are nearly as large as the perikarya of small radial cells. Thus, a growth cone might act as a current source for the action potential arriving along the axon. This view is supported by the finding that, in these early stages of development, the potential reversal occurs not only radially but also tangentially. If one maps the tectal surface responses along the main axis from front to back with micropipettes, a negative potential is recorded in regions covered by fibres; a potential reversal occurs near the growing front of fibres, and in the posterior region, empty of fibres, a positive potential is obtained (Figs. 37, 38).

If growth cones are a possible current source, we have to conclude that although potential reversal is a necessary condition for synaptic activation, it is not sufficient. Thus, additional criteria have to be introduced. One such criterion is the duration of the tectal response compared with that of the pure fibre potential identified with antidromic stimulation (Rager 1976b). Another check was made by applying double shocks at various interstimulus intervals; the amplitude of the positive potential should vary in a similar way to that of the fibre potential if potential reversal is due to growth cones. This does in fact happen on incubation day 9. However, on incubation day 10 the potential decreases with larger interstimulus intervals than one day earlier. With further shortening of the interstimulus interval the amplitude of the response reaches a plateau. Only when the interval is as small as that producing diminution on day 9, does the amplitude further decrease (Rager et al., to be published).

The information obtained from these double-shock experiments combined with the current-source-density analysis indicates that synaptic transmission is already

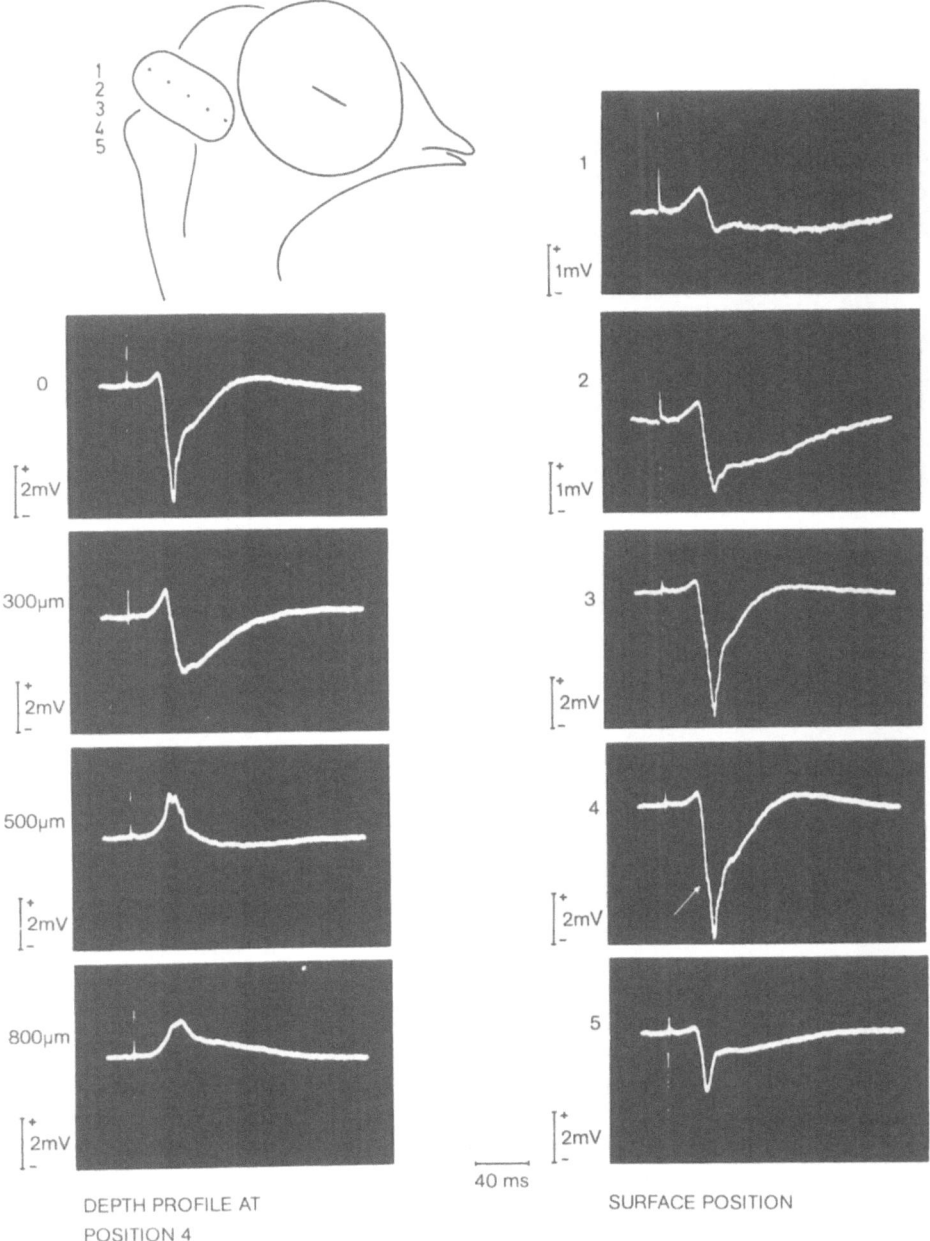

1

+
1mV
-

0

+
2mV
-

300µm

+
2mV
-

500µm

+
2mV
-

800µm

+
2mV
-

2

+
1mV
-

3

+
2mV
-

4

+
2mV
-

5

+
2mV
-

40 ms

DEPTH PROFILE AT
POSITION 4

SURFACE POSITION

Fig. 37a, b. The spread of field potentials in the optic tectum on incubation day 12 (stage 38) mapped radially (a) and tangentially (b). The *arrow* points to a kink in the depolarization phase

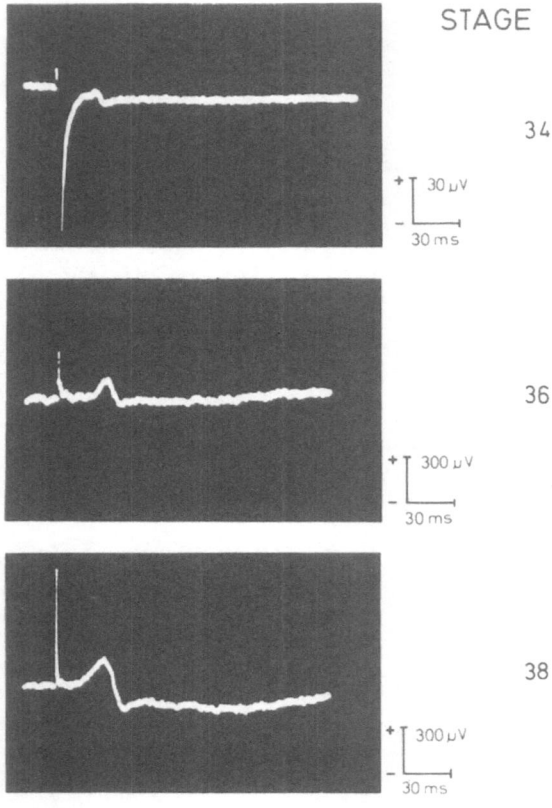

STAGE

34

Fig. 38. Evoked potentials recorded from tectal sites which were not yet reached by the growing front of fibres. Time constant of the amplifier: 100 ms

+ ⊤ 30 μV
− ⌊
30 ms

36

+ ⊤ 300 μV
− ⌊
30 ms

38

+ ⊤ 300 μV
− ⌊
30 ms

effective on incubation day 10. This is consistent with our morphological observations (Sect. 3) and is one day earlier than found previously with other techniques (Rager 1976b).

4.3 Structural and Functional Development

The development of ganglion cell axons, the invasion of the optic tectum by retinal fibres, and the formation of retinotectal synapses were investigated with both morphological and physiological techniques. It turned out that axons, synapses and dendrites start to function as soon as they can be recognized as such from a structural point of view. The reverse is also true; we can draw certain conclusions concerning the structural development of the tectal cortex and its afferent systems from the appearance of certain functional phenomena, such as potential fields in the optic tectum after optic nerve stimulation. Morphological and physiological aspects are not only correlated, they are both important in searching for developmental dynamics underlying the phenomena observed. Dynamic aspects of development will be the main theme of the next two sections.

5 The Formation of the Retinotopic Map

5.1 Retinotopy and Order

5.1.1 The Existence of the Retinotopic Map

It has been shown by electrophysiological mapping and by degeneration experiments in the pigeon that the retina is represented on the optic tectum in such a way that neighbouring retinal fields project to neighbouring tectal sites (Hamdi and Whitteridge 1954; McGill et al. 1966; Clarke and Whitteridge 1976). There is substantial evidence that the retinotectal projection in the chicken is similar to that in the pigeon. This means that there are retinotopy and order. Since "terms such as order, randomness and specificity have been commonly uses . . . and . . . each author is using a privat language and the terms that he uses may mean one thing to him and another to someone else" (Gaze and Keating 1972), I should like to present the following definitions of retinotopy and of an ordered map.

5.1.2 Definition of Retinotopy and Order

It is not yet known whether the projection of retinal ganglion cells to thalamic and hypothalamic nuclei is made directly or via collaterals. In addition, the number of fibres projecting to these nuclei is relatively small. For these two reasons we neglect these connections in the following definition of the retinotectal map.

Neighbouring ganglion cells (x_1, x_2, x_3) project to neighbouring entrance points (y_1, y_2, y_3) at the tectal cortex in a one-to-one way (injective map) (Fig. 39). In addition, each element y ϵ Y has at least one x ϵ X as pre-image (surjective map). Therefore, this map as a whole is bijective. Since X (retina) and Y (optic tectum) differ in structure and size and a field of a given size in the centre of the retina projects to a larger field in the optic tectum than a peripheral one which is due to its higher density of ganglion cells, we consider X and Y as topological spaces. The map $f: X \to Y$ is, therefore, homeomorphic.

However, the homeomorphic map from individual ganglion cells x_i to individual entrance points y_i at the tectal cortex does not exist in reality. Therefore, we have to ask for its actual elementary units. Since the precision of topography in the retinotectal projection of the chick is not yet determined, we have to look for a suitable paradigm. Due to important structural similarities in the organization of the afferent fibre system we may assume that the precision of the retinotectal map is in the same order of magnitude as found in the primary cortical projection fields of mammals. In the visual cortex of mammals functional spatial subunits were found having a diameter of about 2 to 3 mm. The position of receptive fields of cortical cells responding to stimulation at a given retinal point can vary randomly within such a cortical cylinder. Only if the recording electrode is moved a distance equal to the diameter of such a subunit, a new set of receptive fields can be found responding to stimulation of a neighbouring retinal field. On the level of these functional subunits a continuous progression of the retinotopic projection is found (Creutzfeldt et al. 1974; Hubel and Wiesel 1974; Albus 1975). A similar overlap of receptive fields

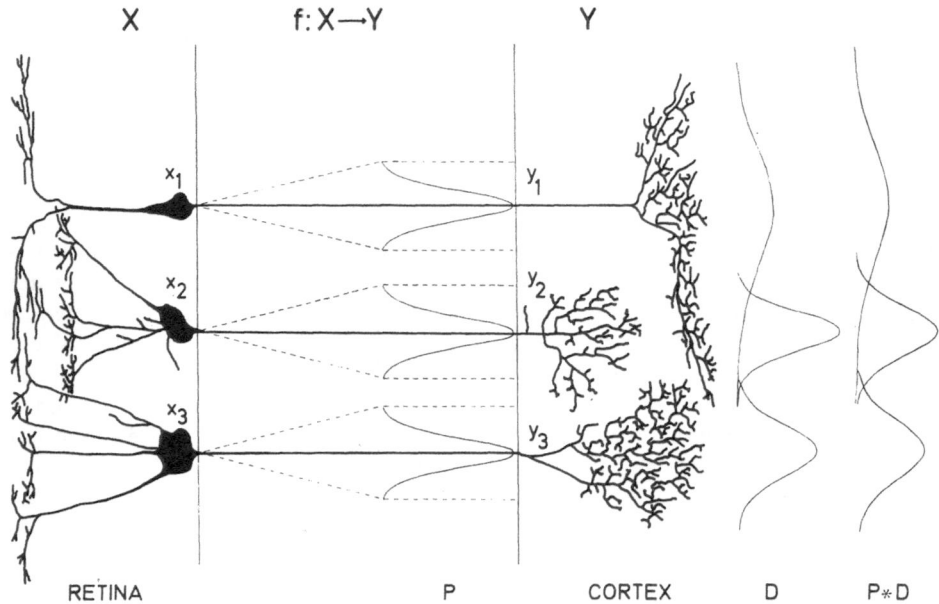

Fig. 39. Schematic diagram to illustrate the definition of retinotopy used in the text. On the left ganglion cell perikarya and dendrites are drawn according to Golgi preparations of Ramón y Cajal (1893, 1911); they are elements of the set X. On the right the entrance points of their axons in the optic tectum being elements of the set Y are drawn together with various forms of axonal arborization. During development, however, a retinal fibre x_i may enter the tectal cortex in a given neighbourhood of y_i with a certain probability symbolized here by the normal distribution function P. In addition, the topography is blurred by the spread of the axonal arbor which can be symbolized by the probability density function D. The curves on the right hand side result from the convolution $P * D$ and represent the final precision of topography. As a result $f: X \rightarrow Y$ is not a point-to-point map but a field-to-field map

seems to exist in the somatosensory cortex of the owl monkey; it reduces the precision of the topography to cortical surface areas of approximately 1—1.5 mm in diameter which "contain all neurons with receptive fields overlapping a particular receptive field any place on the body surface" (Sur et al. 1979). Therefore, we may designate these functional spatial subunits as the elements of the homeomorphic map. Thus, although our considerations start from mathematical concepts, the definition also accounts for biological facts.

How is this homeomorphic or retinotopic map built up during development? There are two contradictory possibilities of explanation: Either the structural arrangement of growing fibres is sufficient to impress retinotopy onto the tectal cortex or it is not; in the latter case additional mechanisms like chemoaffinity would be required to sort fibres out. We shall discuss first the chemoaffinity hypothesis and then analyse the structure of the developing retinotectal projection. Finally, we shall ask whether the conditions generated by the structural development of the whole system would give us an alternative to the chemoaffinity hypothesis.

5.2 The Chemoaffinity Hypothesis

The phenomenon of topologically ordered projections gave rise to quite a number of hypotheses succeeding each other in a dialectic way. The chemotropism hypothesis (e.g. Ramón y Cajal 1911) was followed by hypotheses explaining the fibre growth and the formation of synaptic connections on predominantly mechanical grounds (e.g. contact guidance, resonance principle (Weiss 1928, 1936)). Then Sperry (1963) came to the conclusion that mechanical interpretations were not sufficient to explain the selectivity of nerve connections. He postulated "that the patterning of synaptic connections in the nerve centres . . . must be handled . . . by the growth mechanism directly, independently of function, and with very strict selectivity governing synaptic formation from the beginning." (p. 703). During development and growth "specific cytochemical affinities" arise which would have to be understood as "individual identification tags, presumably cytochemical in nature" (p. 703–704). These chemical labels would enable retinal ganglion cells to recognize their partners in the optic tectum because of complementary molecules. Thus, a retinotopic projection can develop. Meanwhile the chemoaffinity hypothesis formulated by Sperry (1943, 1963) has been changed and modified in many ways. At present it seems to be the hypothesis most widely used to explain the development of topologically ordered connections.

Difficulties and objections with respect to the original version of the chemoaffinity hypothesis are now arising from theoretical considerations. Firstly, according to Sperry (1963) it is "a necessary conclusion" from the observation of selectivity of neuronal connections "that the cells and fibres . . . must carry some kind of individual identification tags, presumably cytochemical in nature." The necessity of this conclusion has not yet been proven (Gaze and Keating 1972). Secondly, a scientific theory or hypothesis should be refutable or falsifiable (cf. Popper 1959) either directly or in one of its consequences or predictions. This requirement is not fulfilled by the chemoaffinity hypothesis. This is recognized by Hunt and Jacobson (1974, p. 242): "Some aspects of neuronal specificity may remain refractory to direct physicochemical analysis for some time." Therefore, neuronal specificity is usually tested indirectly by some surgical interference. But here we meet another objection. "Sperry's theory . . . is a theory of normal development . . . and not a theory of the behaviour of neurones in artificial environments. Thus, surgically created contexts . . . may have only limited relevance for the theory." (Hunt and Jacobson 1974, p. 220) How then can the hypothesis be tested?

Difficulties also arise from experimental results. Constantine-Paton and Capranica (1976a, b) have shown that retinal fibres do not necessarily contact tectal cells. Although the eyes were transplanted to the position normally occupied by the ear, which is near to the optic tectum, retinal fibres grew down to the medulla and to the spinal cord. Since the dorso-caudal trajectory of optic nerve fibres was "very similar to the growth trajectory in the normal diencephalon", it was concluded "that the dorso-caudal growth relative to the three major axes of the neural tube is an inherent property of optic nerve fibres" and not imposed by polarity cues in the diencephalon or in the optic tectum. However, if a third eye is grafted onto the spinal cord of Xenopus larvae, fibres do not show a dorso-caudal gradient but grow rostrally and terminate near the nucleus tractus solitarii. Some fibres also grow caudally (Giorgi and Van der Loos 1978). These phenomena indicate that chemoaffinity does not prevent retinal fibres from terminating in other regions of the central nervous system and that the direction of growth is not inherent in retinal fibres.

In addititon, there is now increasing evidence that tectal markers may not be necessary for topographic ordering of retinal fibres and may not even be present prior to the innervation by optic fibres (Schmidt 1977; Cook and Horder 1977; Constantine-Paton and Law 1978). Thus, new models have been developed emphasizing the importance of the positions of retinal ganglion cells (Hope et al. 1976), of optic fibre guidance and of direct fibre-fibre interactions (Cook and Horder 1977). In the model of Willshaw and von der Malsburg (1979) and von der Malsburg and Willshaw (1977) the two sheets of retinal and tectal cells do not acquire their markers independently and before any connections are made as postulated by Sperry (1963); the tectal markers are induced by retinal fibres.

The consequence of these considerations is that we should analyse the development of the individual elements cooperating in the formation of the retinotectal connection in greater detail. The following questions are, therefore, put forward: (1) How does the intraretinal organization develop? (2) How are fibres arranged in the optic fibre pathway of the developing and of the adult animal? We cannot be satisfied with the fact that retinotopy or homeomorphism are found at the optic tectum. In order to get some insight into possible developmental mechanisms, we should study each intermediate position: the entrance into the optic stalk, the optic nerve, the chiasm and the optic tract. (3) How does the optic tectum develop? Which factors determine the origin and the reference frame of the retinotectal map?

5.3 The Organization of Fibres in the Visual Pathway

5.3.1 Minimal Conditions of Order

There is a continuing debate whether the fibres in the optic pathway of various species are orderly or even retinotopically arranged or not. This debate is partly due to the lack of a generally accepted definition of a retinotopic organisation. We, therefore, start from the definition given in Sect. 5.1 and ask what are the minimal conditions which have to be fulfilled to establish a retinotopic map with the precision of a field-to-field projection. The minimal conditions are the following two. Firstly, each individual fibre x_i (Fig. 39) may vary its position in the fibre pathway considerably since it is only required that it can be found in a sufficiently small neighbourhood of the entrance point y_i with a sufficiently high probability. This probability P can be expressed by distribution functions. Furthermore, the tangential spread of the axonal arborization can be represented by two-dimensional probability density functions D showing a continuous overlap of neighbouring entrance points. The convolution $P * D$ should give the actual precision of the map. Secondly, transformations may occur at certain places in the fibre pathway. Again, it is not required that the transformation rules are followed with a point-to-point precision; a field-to-field mapping would be enough. Therefore, although a high degree of order should exist, it is not necessary that this order is strictly retinotopic at every level in the fibre pathway.

5.3.2 Retinotopy in the Optic Fibre Layer of the Retina

Retinal ganglion cells are first generated at the centre of the retina (Sect. 3). With time, generation of ganglion cells progresses to peripheral regions. During migration from the ventricular zone towards the inner limiting membrane, each ganglion cell forms an axon which immediately grows towards the optic fissure. The first ganglion cell axons enter the optic stalk on incubation day 3. We have prepared Golgi-impregnated whole mounts of the retina at various stages of development. This technique allows us to see perikarya as well as axons of ganglion cells at defined locations and to follow these axons over quite long distances. The axons of neighbouring cells join together, form fascicles, and run parallel to each other towards the optic fissure by the shortest possible route (Figs. 40, 41). However, at the beginning of fibre outgrowth, the fibre arrangement is not as regular as that seen at later stages. Fibres are progressively straightened out. This phenomenon may reflect the stretching of the growing retina (Goldberg and Coulombre 1972). This regular pattern of fibre outgrowth is not confined to the retinal centre, but can be found also in more peripheral parts of the retina. Only in the extreme periphery, are fibres found to grow parallel to the retinal circumference. These circumferential fibres enter the more peripheral end of the optic fissure (Goldberg and Coulombre 1972).

It is fortunate that the tips of growing axons can easily be identified by their growth cones. Since central ganglion cells are generated earlier than peripheral ones, their axons should reach any given position in the fibre pathway first. Thus, if we cut the fascicles of ganglion cell fibres at an appropriate time and position, we should see the thin axons of central ganglion cells which have already passed beyond the site of the cross-section together with growth cones of axons from more peripheral cells which have just arrived. This can be seen in Fig. 9, where the central retinal area is cut perpendicularly to the course of the fibre fascicles. It is clear that the distal half of fibre bundles consists mainly of thin axons, whereas the proximal part, which is near the inner limiting membrane, consists mainly of growing tips. This indicates that peripheral fibres are located predominantly in the proximal part of retinal fascicles, and this is consistent with our Golgi material and with the finding of Goldberg and Coulombre (1972) obtained from silver-stained retinal whole mounts. Thus the neighbourhood relations of ganglion cell perikarya are represented with a high probability in both the tangential and the radial axis of the fibre layer of the retina itself.

5.3.3 The Transformation of the Fibre Organization at the Entrance of the Optic Nerve

We can localize central and peripheral fibres in the developing optic nerve in the same way. Fig. 23 shows a cross-section of the optic nerve on incubation day 5 having several characteristics of the optic stalk. The dorsal part is still filled with undifferentiated neuroepithelial cells. The optic stalk lumen is still open. The anterior, ventral and posterior parts show small pale areas. If we magnify these areas with the electron microscope, we see that the fibre bundles located in the central part of the fibre area contain mainly thin axons, whereas bundles located in the nerve's periphery contain mainly growth cones. This is shown in the electron micrographs of Fig. 26 and 27 which were taken from the same ultrathin section and at the same magnification. In

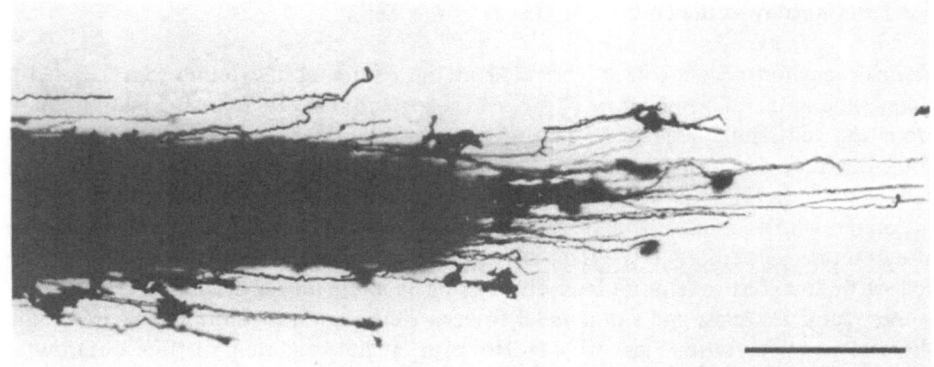

Fig. 40. Golgi-impregnated whole mount of the retina on incubation day 7. The central retina is shown. The choroid fissure is near the left border. *Bar* 70 μm

other words, central retinal fibres tend to be located in the centre of the fibre area, whereas peripheral retinal fibres are located mainly in the periphery. We cut the whole optic nerve successively at small intervals (50–100 μm) and always found this organization of central and peripheral fibres, regardless of distance from the retina.

If we study the ingrowth of retinal fibres into the optic stalk at successive stages of development, we see that the first fibres to arrive occupy a thin rim next to the ventral pial border on incubation day 4. Thus, the central retinal area, which is presumably circular, is transformed into a crescent in the optic stalk. One day later (incubation day 5) a new larger crescent of fibres is added peripherally. Thus the position of the earlier-formed fibres becomes more central in the nerve cross-section. This pattern of invasion is continued during the next few days until the whole nerve is filled with fibres. Since we know that there is retinotopy in the optic nerve of the adult animal, at least as far as quadrants and central and peripheral fibres are concerned (Sect. 5.3.4), the question arises: How is this early situation transformed into that of the older animal? The individual developmental steps have not yet been analysed fully, but the transformation could be explained as follows: Firstly the "rings" of incremental growth of retinal ganglion cells are transformed into crescents of fibres in the optic nerve. This process would be analogous to what has been observed in cichlid fishes (Scholes 1979). It is similar in that peripheral fibres are located vitreally and enter the papilla ventrally, it is different in that we have to postulate that in the chick not the ventral, but the dorso-nasal fibres are at the two ends of each crescent. This postulate is based on the following reasons. First, if the dorso-nasal fibres were located in the centre of each crescent, they would have to cross to reach their final dorsal position in the nerve. Second, it can be observed in silver-stained retinal whole mounts (Goldberg and Coulombre 1972) and in Golgi-Cox impregnated eye-cups that dorso-nasal fibres form an arcuate pattern around the central area which is slightly elevated during this period of development. The dorso-nasal split of the ring may occur due to this structural speciality. Since the distance from the papilla is larger for dorso-nasal fibres than for others of the same incremental ring, it is likely that dorso-nasal fibres join the crescent at its two ends. Later on, these crescents are compressed again by the invasion of new fibre masses. In addition, the neuroepithelial cells, which originally separate the two ends of the split ring, eventual-

46

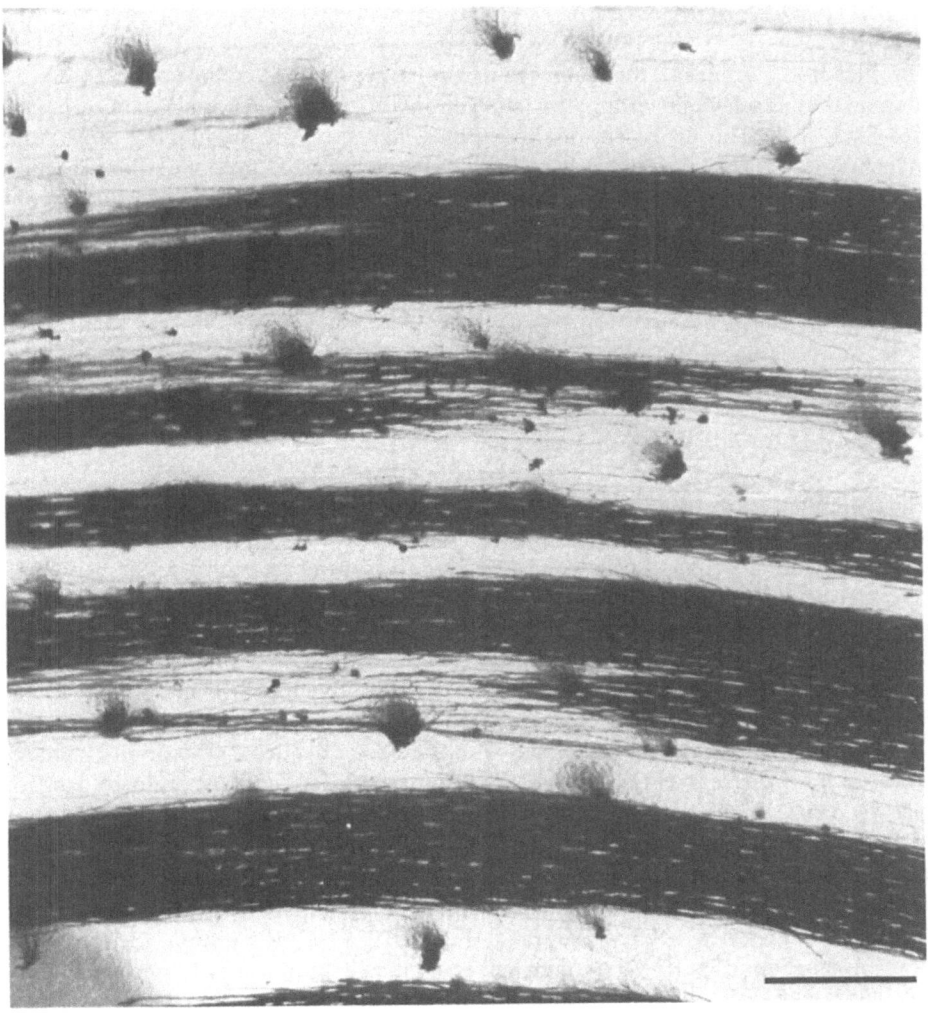

Fig. 41. Golgi-impregnated whole mount of the retina on incubation day 7 showing the parallel course of individual fibres and of fibre fascicles. *Bar* 130 μm

ly degenerate and disappear (Fig. 25) or are transformed into glial cells. Thus, at the final stage the topographical organization of retinal fibres matches that of the corresponding perikarya. (A further analysis of these events is given in Rager 1980).

5.3.4 The Fibre Organization in the Middle of the Optic Nerve

It is well known that the optic nerve of mammals is organized retinotopically in its middle portion (Henschen 1910; Polyak 1957). Does a similar organization exist in the chick optic nerve? We tried to answer this question with morphometric techniques using myelination, fibre density, and fibre diameters as parameters.

5.3.4.1 The Onset of Myelination

The first fibres to invade the tectal cellular layers are those from the central retina. As soon as they reach their termination site (on incubation day 8), they start to branch (see Sect. 5.4). The diameters of these fibres increase with the number of terminal branches (see Sect. 6), and when they reach about 0.8 μm, the fibres become myelinated (Sect. 3). Thus, it is possible to identify the fibres from the central retina as those which become myelinated first. A morphometric analysis of electron micrographs taken from randomly selected areas of optic nerve cross-sections (see Sect. 2) revealed that the first myelinated fibres, i.e. the central retinal fibres, were located in the central core of the nerve on incubation days 15 and 16, which means that there is retinotopy as far as the representation of the central and peripheral retina is concerned.

5.3.4.2 The Distribution of Fibre Density

The fibre density distribution is a useful parameter in testing for retinotopy only if we know more about the distribution of ganglion cell densities and sizes in the retina. The density of retinal ganglion cells decreases from the central area towards the periphery. This can be recognised qualitatively from the decreasing thickness of the ganglion cell layer, but it has also been shown quantitatively (Polyak 1957; Van Buren 1963; Stone 1965; Hebel 1976). In addition, the average sizes of perikarya and dendritic fields increase towards the periphery (Binggeli and Paule 1969; Boycott and Wässle 1974). These morphological findings are supported by electrophysiological results; the size of receptive fields is closely correlated with the size of dendritic arbours and increases towards the periphery (Fischer 1973; the contribution of the structure of ganglion cell dendrites to the properties of receptive fields is discussed by Creutzfeldt et al. 1970). However, the decrease of ganglion cell density and the increase of cell size towards peripheral retinal regions is not simply a function of distance from the central retinal area. The central area of many animals is deformed to a streak-like pattern (Stone 1965; Hughes 1975; Vaney and Hughes 1976). In the pigeon, the isodensity lines of ganglion cells show two maxima: one is located in the central area (monocular lateral visual field), the other is located in the dorso-temporal quadrant at an eccentricity of 40°–50° (binocular frontal or pecking field) (Galifret 1968, Binggeli and Paule 1969). It is highly probable that the chicken retina shows a similar structure (Galifret 1968). We have serially sectioned whole eyes of developing as well as adult animals. A qualitative analysis indicates that the distribution of ganglion cells is quite similar to that found in the pigeon. In addition, the largest ganglion cell perikarya are located in the ventro-nasal periphery. It is well established that in the adult animal axon diameters reflect the sizes of the corresponding perikarya (Stone and Holländer 1971; Boycott and Wässle 1974). Thus, if there is retinotopy in the optic nerve, we should expect to find the largest fibre diameters and, therefore, the lowest fibre density in the ventro-nasal periphery of the nerve.

Binggeli and Paule (1969) divided the adult pigeon's optic nerve into quadrants and into central and peripheral regions to test for differences in fibre density. "In all instances, there was no significant difference (at the P 0.05 level) in the fibre density between various locations in the nerve." In contrast to these results we found in the chicken that there are considerable regional differences in fibre density and fibre

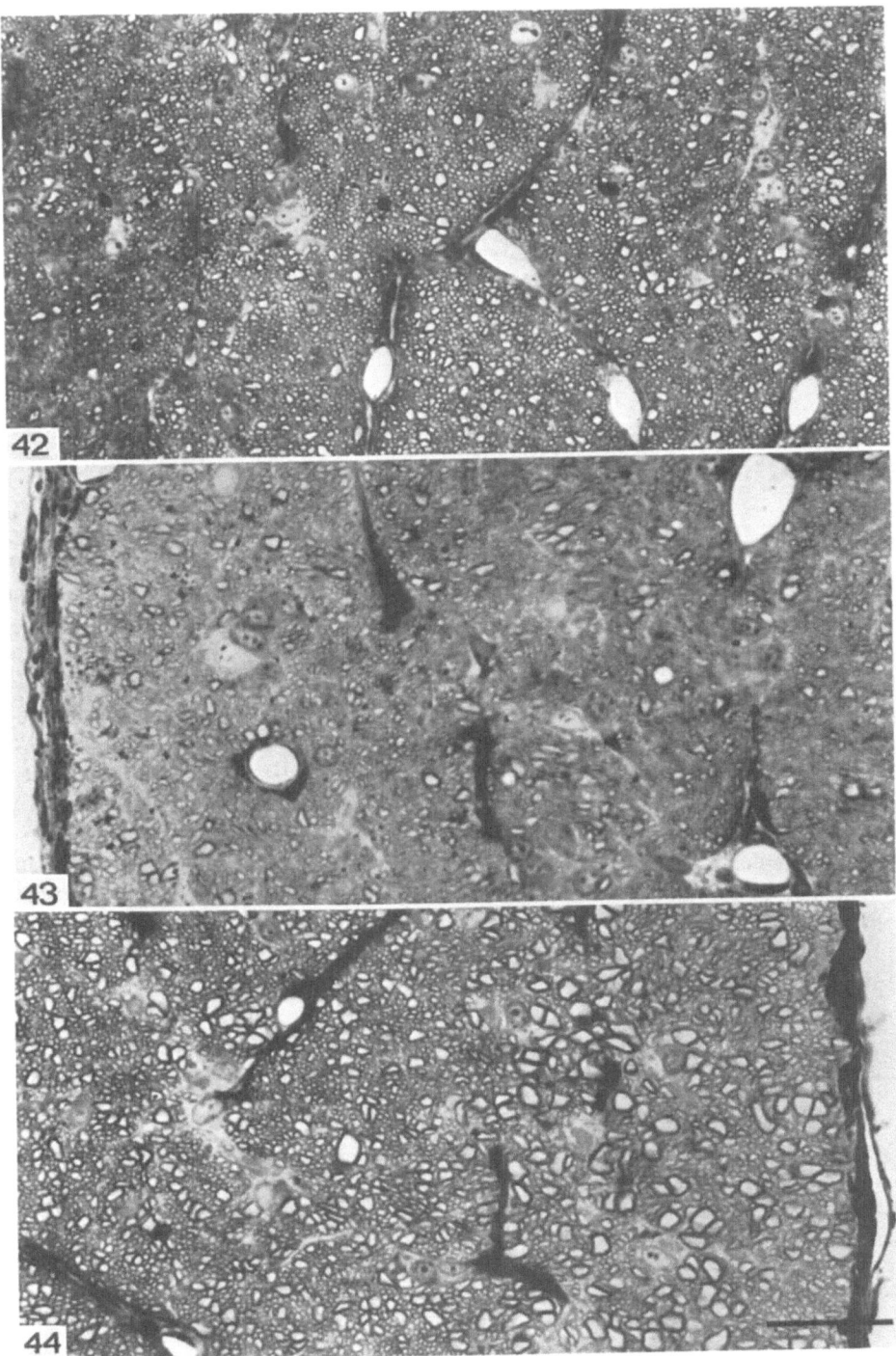

Figs. 42–44. Representative small areas of the cross-section of the optic nerve of chicken 104 days after hatching taken from the same semithin section with the same magnification. It is clear that the fibre diameters in the ventro-nasal periphery (Fig. 44) are larger than in the central core (Fig. 42) and in the dorso-temporal periphery (Fig. 43). *Bar* 25 μm

49

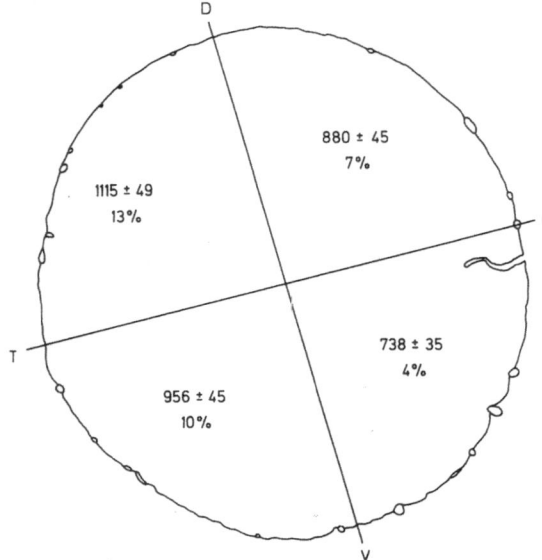

D

880 ± 45
7%

1115 ± 49
13%

N

T

738 ± 35
4%

956 ± 45
10%

V

Fig. 45. The cross-section of the optic nerve of a 104-day-old chicken subdivided into standard quadrants. Fibre density (number of fibres/1000 μm^2) and percentage of unmyelinated fibres are indicated in each quadrant. *D*: dorsal, *T*: temporal, *V*: ventral, *N*: nasal

diameters even in semithin sections of optic nerves of the adult chicken. This can be seen in Fig. 42, 43 and 44 (the three micrographs were taken at the same magnification and from the same semithin section of a 104-day-old chicken), where the central area, the dorso-temporal periphery, and the ventro-nasal periphery are shown respectively. It is clear that fibres in the ventro-nasal periphery reach much larger diameters than in the other two regions. The same nerve has been analysed quantitatively under the electron microscope. Sixty micrographs were taken per quadrant so that a total of 240 micrographs was taken from the whole nerve. Altogether 11,300 fibres were counted and measured at a total magnification of × 28,000. The results are summarized in Fig. 45. The fibre density (number of fibres per 1000 μm^2) reaches a maximal value of 1149 (standard deviation ± 49) in the dorso-temporal quadrant, a minimal value in the ventro-nasal (738 ± 33), and intermediate values in the ventro-temporal (956 ± 45) and dorso-nasal (880 ± 45) quadrants. Thus, fibre density is 34% less in the ventro-nasal than in the dorso-temporal quadrant. The percentage of unmyelinated fibres varies in a similar way: we find 13% dorso-temporally, 10% ventro-temporally, 4% ventro-nasally, and 7% dorso-nasally. The mean fibre density

Table 3. Fibre densities in the quadrants of the optic nerve

Quadrant	Degrees of freedom	Density Core/Ring	t_0[a]	Significance level
Dorso-temporal	57	1.09	1.395	< 0.1
Ventro-temporal	56	1.19	1.818	< 0.05
Ventro-nasal	56	1.2	2.125	< 0.025
Dorso-nasal	57	1.26	2.084	< 0.025

[a] Defined according to Kreyszig (1968)

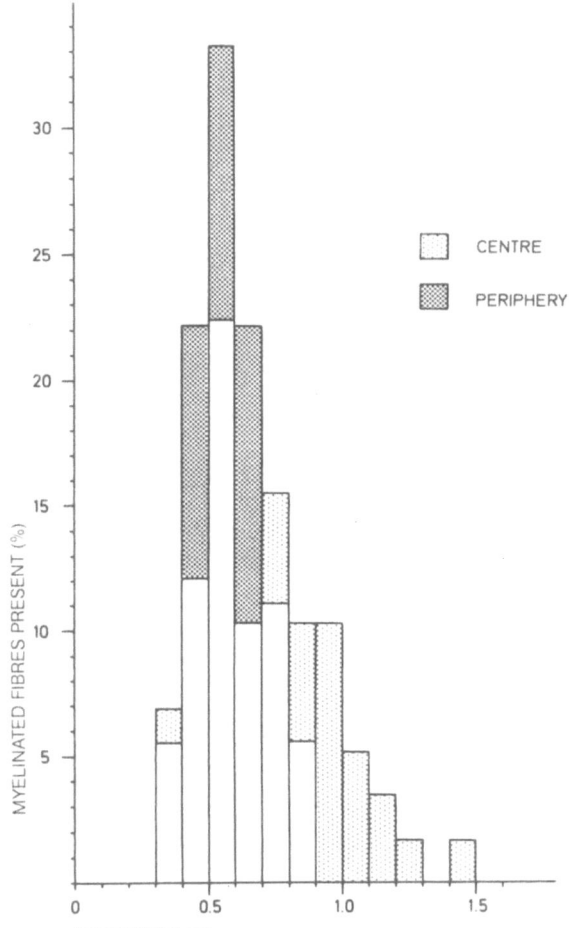

Fig. 46. Distribution of myelinated fibres in the central core and in the periphery on incubation day 18 (stage 44). Fibres having diameters $D > 0.9$ μm can be found only in the central core. Smaller diameters are found predominantly in the periphery

of the nerve's periphery is 818 ± 38, whereas that of the nerve's core is 1006 ± 26 (the core is defined as the central region whose area is equal to the half of the whole cross sectional area: see Fig. 1 and 2). This difference is significant at the $P < 0.001$ level. Although fibre density is higher in the core than in the ring in each quadrant, the difference is smallest in the dorso-temporal and greatest in the ventro-nasal quadrant (Table 3). Thus the fibre density distribution in the optic nerve reflects the ganglion cell density distribution in the pigeon retina (Galifret 1968; Binggeli and Paule 1969) which seems to be similar to that observed in our serial sections of the chicken retina as well as in whole mounts. We find relatively small ganglion cells at a high packing density in the dorso-temporal region as well as in the central retinal area. They are represented by relatively small and densely packed fibres located in the corresponding regions of the optic nerve. This is further evidence for the presence of retinotopy in the optic nerve.

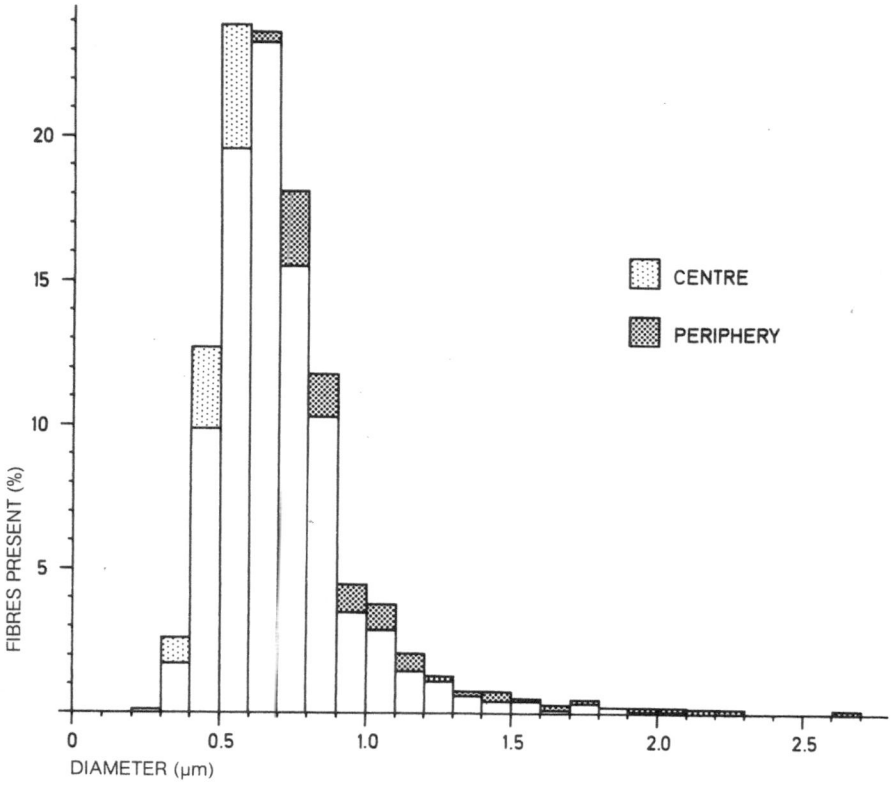

Fig. 47. Fibre distribution in the central core and in the periphery of the nerve 104 days after hatching

5.3.4.3 The Distribution of Fibre Diameters

Fibre density is a relatively crude parameter to test for retinotopy. It represents not only fibres, but also glial cells and blood vessels. For example, during the early phase of myelination we found that glial cells are present in a higher density in the ventro-nasal than in the dorso-temporal quadrant. This can be understood from the fact that glia proliferation is intensified in regions where myelination is taking place. In order to see whether fibre density decrease really does reflect an increase of axon diameters, we measured the diameters of myelinated fibres at various stages of development from incubation day 15 until 100 days after hatching. Indeed, it was found that the diameter distribution was inversely proportional to the fibre density distribution. We shall consider at first the diameters of central versus peripheral fibres and then compare the fibre spectra in the four quadrants.

At the beginning of myelination myelinated fibres with small diameters predominate in the periphery, and those with large diameters in the centre; the largest fibres can be found in the centre only. The reason for that is that central retinal fibres mature first (Fig. 46). This relation is completely reversed in the adult animal: Among the central fibres small diameter classes predominate, fibres with diam-

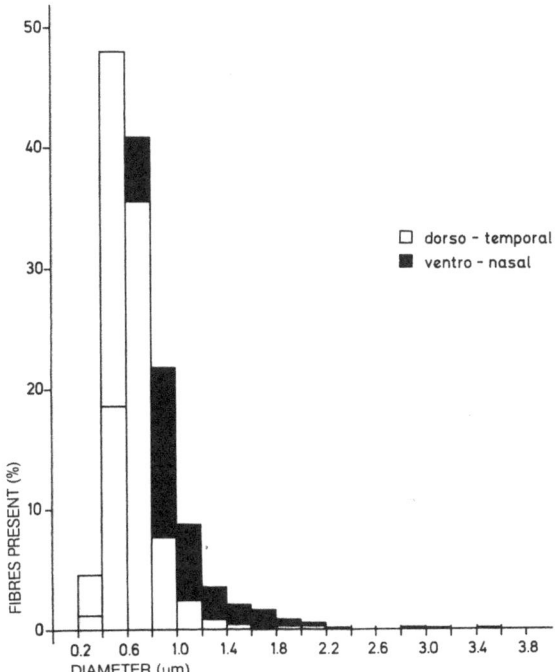

Fig. 48. Fibre distribution in the ventro-nasal and dorso-temporal quadrant in the nerve of a 104-day-old chicken

eters larger than 0.8 μm are found mainly in the periphery; fibres with diameters larger than 2 μm can be found only in the periphery (Fig. 47). This is presumably related to the fact that peripheral retinal ganglion cells are generally larger than central ones even though they start to mature later in development.

We have also analysed fibre diameters in the four quadrants of the adult animal. In Fig. 48 the distributions of fibre diameters in the dorso-temporal and ventro-nasal quadrants are compared. Small diameter fibres predominate in the dorso-temporal, fibres larger than 0.6 μm in the ventro-nasal quadrant. The very large fibres can be found only in the peripheral part of the ventro-nasal quadrant. Since the largest ganglion cells are located in the ventro-nasal periphery of the retina, and the largest fibres are found in the ventro-nasal periphery of the nerve, and since the size of peri-karya corresponds to the size of fibre diameters, we again have support for the con-clusion that retinotopy is present in the middle of the optic nerve.

5.3.5 The Course of Fibres in the Optic Nerve, Chiasm and Tract

Whole brains were impregnated with the Golgi-rapid technique, then cut parallel to the optic nerve (Fig. 49). This technique enabled us to follow individual fibres over quite long distances. Most of the fibres run parallel to each other even across the chiasm (see also the sagittal section in Fig. 50); only very few fibres change their relative position.

Optic tract fibres were traced in the same way. It was found that a number of fibres cross the main stream shortly behind the chiasm. It could be that these are fibres terminating in thalamic nuclei or that another transformation of the map

53

Fig. 49. The two optic nerves, the chiasm, and the anterior part of the tract on incubation day 7. The Golgi-impregnated brain was cut tangentially to the fibre course. *Bar* 80 μm

occurs. The majority of fibres, however, show a high degree of order as they grow towards the optic tectum (Fig. 51).

Since fibres fan out as they approach the optic tectum, whole mounts of the optic tract are more suitable than cross-sections in this region. Silver-stained whole mounts clearly show that optic tract fibres maintain their neighbourhood relations (Goldberg 1972). If horseradish peroxidase is injected into a small region of the stratum opticum

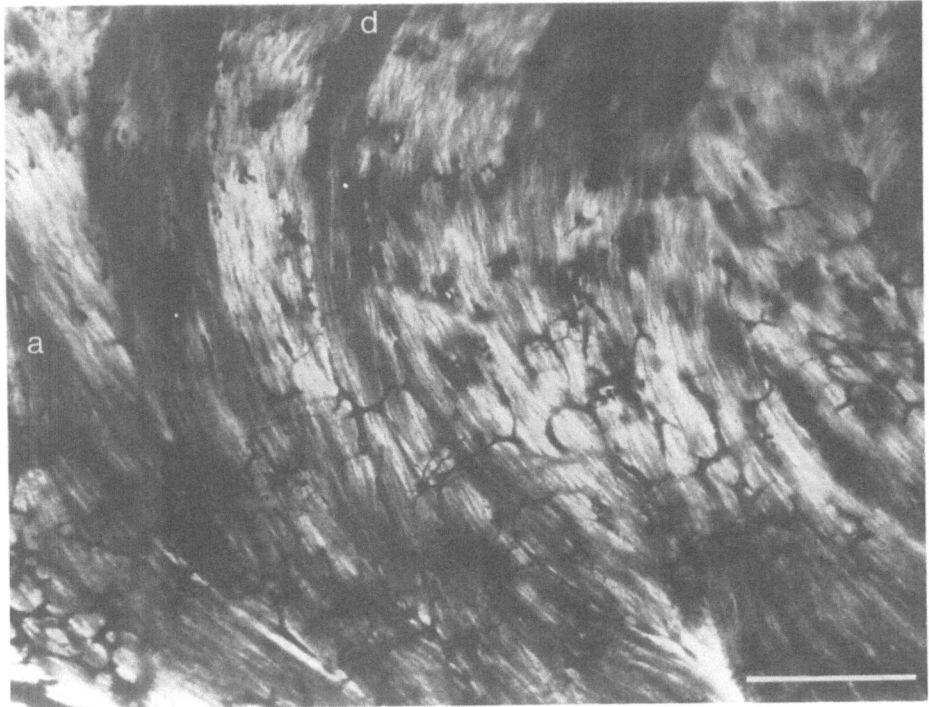

Fig. 50. The chiasm of a 13-day-old embryo cut in a sagittal plane. The Golgi stain shows that fibres have a high degree of order in each of these interdigitating bundles. *a*: anterior, *d*: dorsal. *Bar* 160 μm

(SO) in the optic tectum, only small bundles of fibres are stained. These bundles could be followed up to the eye. It was found that fibres containing the reaction product stayed together in these small bundles in the SO as well as in the optic tract (Horder et al. 1979; Rager et al. to be published).

From these results we may conclude that a high degree of order is present in the optic fibre layer of the retina, in the optic nerve, in the chiasm, in the optic tract, and in the stratum opticum of the optic tectum. They show in addition that the retinotopy of the optic fibre layer in the retina is transformed to the retinotopy of the optic nerve. It is not yet known where the second transformation occurs which should be equivalent to a reflection in the horizontal meridian. Whether this occurs right behind the chiasm is now being explored. The HRP containing fibres stay close together in bundles whose cross sectional area is smaller than that of the nerve by two orders of magnitude. Thus, the degree of order found in the visual pathway may fulfill the requirements to establish the homeomorphic map defined above.

5.4 The Origin of the Retinotopic Map

The tectal area to which central retinal fibres project is located not at the rostral pole, but ventrolaterally at the transition from the rostral to the central third of the optic

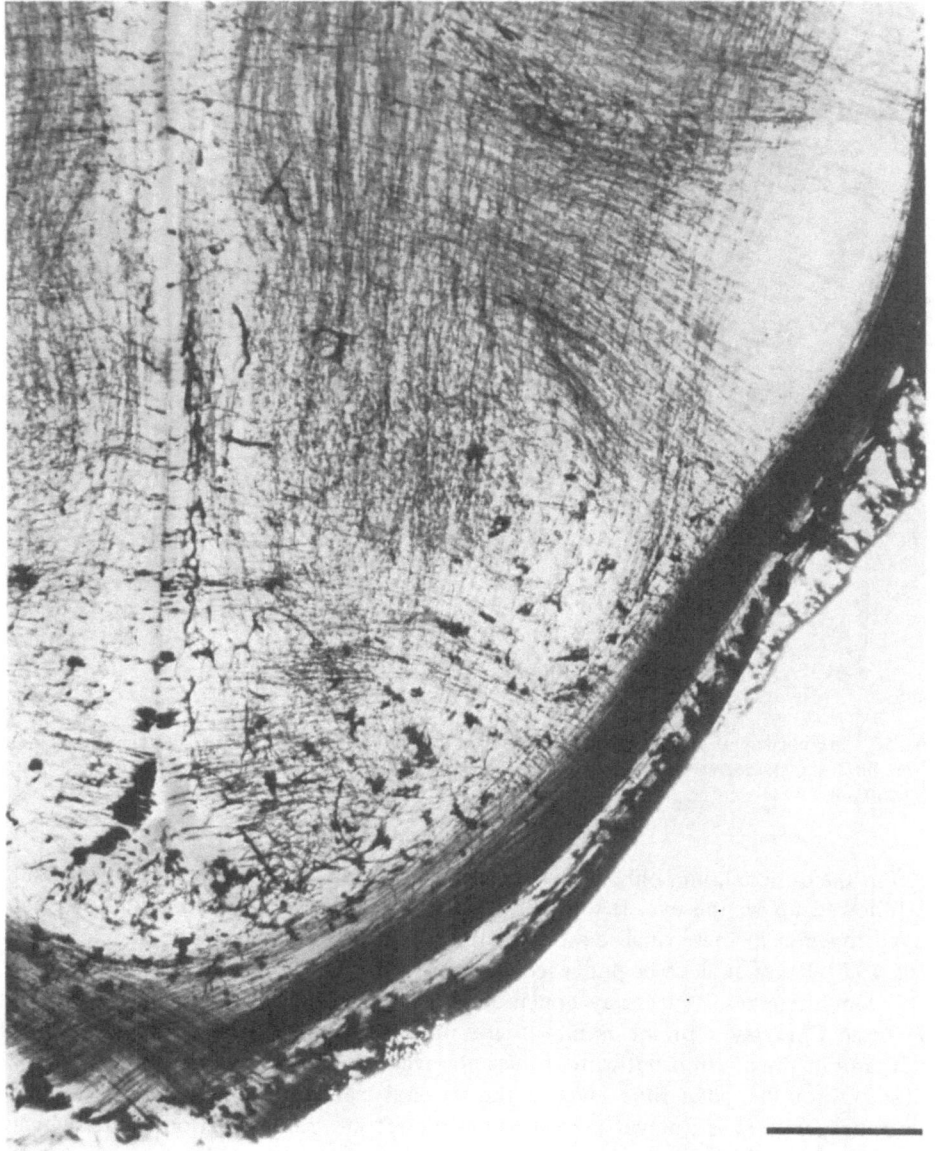

Fig. 51. The optic tract of an 8-day-old embryo is shown in its whole extent from the chiasm to the optic tectum. Note that nearly all of these Golgi-impregnated fibres run parallel to each other. *Bar* 300 μm

tectum (La Vail and Cowan 1971a; Crossland et al. 1974; Rager 1976b). Because this area receives the fibres from the central retina, it is called the central tectal area. Since central retinal fibres are the first which connect to tectal neurons, it forms the origin of the retino-tectal map whereby the origin is understood in the Cartesian sense. Now the question arises: Why do central retinal fibres not contact the tectal cells on the rostral pole, which they meet first during development, i.e. why do they continue to grow until they arrive at the future central area? Or in other words: How is the origin of the retinotectal map determined? Obviously it cannot be determined solely by the arrangement of retinal fibres in the optic nerve.

This problem was investigated recently by Crossland et al. (1975) who used light microscopic autoradiographic techniques. They found that fibres of the first ganglion cells generated, do, in fact, not penetrate the tectum at its rostral end, where they first arrive, but continue to grow over the tectal surface and invade the stratum griseum et fibrosum superficiale (SGFS) only near the central portion of the tectum where they arrive on incubation day 8. With these light microscopic autoradiographic techniques it looks as if fibres do not innervate the cellular tectal layers immediately after arrival; they appear to "wait" at their proper termination site for two days until the conditions become established for their entry. This pattern of fibre growth has been "regarded as strong a priori evidence in support of a chemoaffinity mechanism comparable to that first postulated by Sperry" (Crossland et al. 1975) and has been discussed in Sect. 5.2.

Since light microscopic autoradiography cannot resolve individual fibre branches and might not allow the detection of axons just leaving the stratum opticum (SO), we have studied the ingrowth of retinal axons with the Golgi technique (Rager and von Oeynhausen 1979). For this purpose the brains of 56 embryos at various developmental stages were stained with a modified Golgi-rapid method (see Sect. 2). They were then cut in frontal, horizontal, and sagittal planes and parallel to the incoming fibres at the central tectal area.

We found that on incubation day 6, when the first retinal axons reach the rostral pole of the optic tectum (De Long and Coulombre 1965; Goldberg 1974; Crossland et al. 1975) only two layers superficial to the ventricular zone could be recognized (La Vail and Cowan 1971a). Thus, no appropriate partners are available for arriving fibres to make contact with. Fibres continue to grow over the tectal surface. The formation of a cortical plate does not start before incubation day 7 when pear-shaped radial cells accumulate below the pial surface (Puelles and Bendala 1978). These cells start to sprout dendrites growing as narrow, radially oriented tufts. On incubation day 8 the apices of a number of these dendrites even invade the SO (Fig. 53). In general, at this stage the central region (Fig. 52) stands out with respect to its cytoarchitectural development (Leghissa 1957; Cowan et al. 1968; La Vail and Cowan 1971a, b; Puelles and Bendala 1978), the development of cell types such as stellate and elongated cells (Puelles, personal communication), and the maturation of dendrites of superficially located cells. At about the same time the first retinal fibres have reached this most advanced tectal region (Goldberg 1974; Crossland et al. 1975; Rager 1976a; Fig. 53) which roughly corresponds to the central tectal area in the adult animal. Thus, retinal fibres meet sufficiently mature dendrites of radial tectal cells in the SO for the first time when they arrive at the central tectal area. The location and the extent of this region is indicated in Fig. 52 (arrows). It can be recognized from its thickness and lamination; these both decrease considerably towards the periphery of the tectum.

On incubation day 8, in the central tectal area, we can see not only growing dendritic tips invading the layer of arriving fibres, but also a few ganglion cell axons leaving this fibre layer and invading the SGFS (Fig. 53). The axon shown in Fig. 54 could be traced for quite a long distance through the SO. At the point where it turns down to enter the tectal cortical plate radially it gives off two side branches which continue to grow within the SO. At a distance 20 μm below the optic fibre layer, a

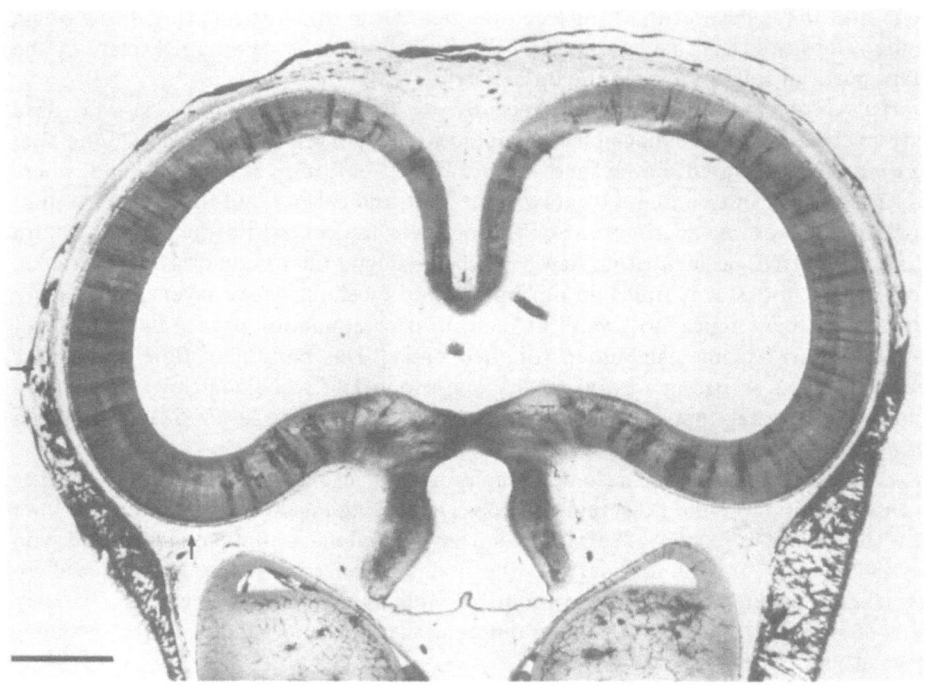

Fig. 52. Tectal hemispheres in a Golgi-impregnated brain of an 8-day-old embryo. The brain is cut parallel to the incoming afferent fibres in the central tectal area which is demarcated by *arrows*. Note the difference in maturation between central and peripheral tectal regions. *Bar* 1 mm. (With permission of Springer-Verlag)

growth cone can be recognized having sheet-like membranous extensions (foliopodia) which are thought to be characteristic of growing ganglion cell axons (Hinds and Hinds 1974; further criteria for the identification of retinal axons are given in Rager and von Oeynhausen 1979). Similar invading axons were found in adjacent sections (Fig. 55) and in other brains of the same developmental age. These observations indicate that there are at least some fibres which do not wait at their termination site for two days. On the contrary, there seems to be a continuous progression from arrival of fibres to their ingrowth and even to the formation of contacts (Sect. 3.4). Perhaps the apparent "waiting", found in other systems, should not be interpreted as a real standstill in fibre growth either. In the case of geniculo-cortical connections fibres arriving from the dorsal lateral geniculate body "may contact migratory neurones passing through the intermediate zone, before they reach the appropriate level in the cortex" (Rakic 1977).

It is much easier to find ingrowing retinal fibres in the central tectal area on incubation day 9 than on day 8. On incubation day 10 most axons ramify extensively in the superficial layer IX and X (designation according to La Vail and Cowan 1971a); other axons first grow to deeper layers before they start branching (Rager and von Oeynhausen 1979). Their terminal branches can even be found in layers as deep as those in the adult animal. From incubation day 10 onwards synaptic contacts can be observed in the superficial laminae of the SGFS with the electron microscope

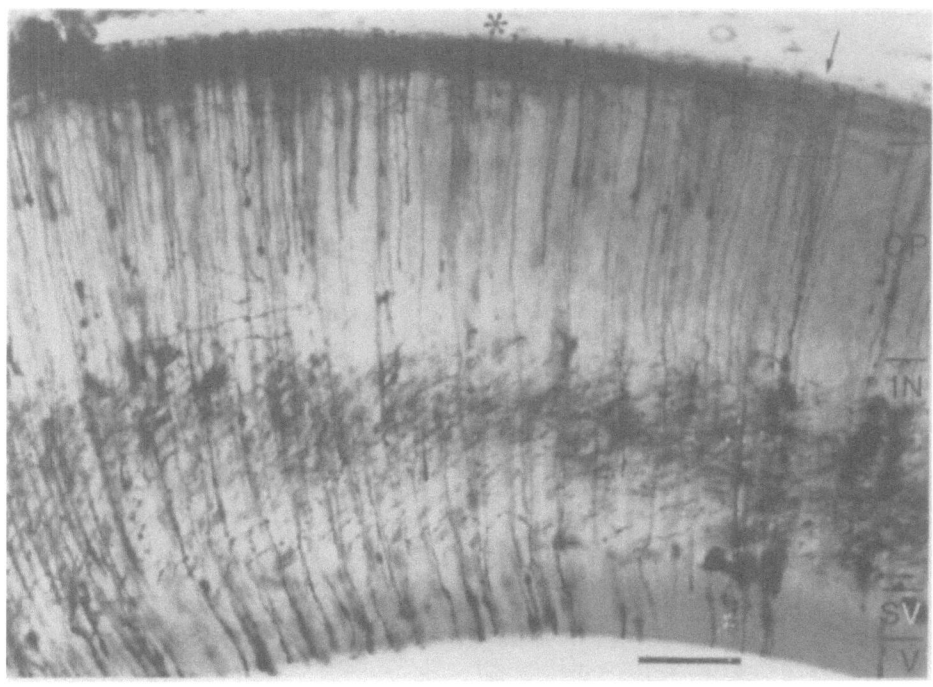

Fig. 53. Central tectal area on incubation day 8 stained with the Golgi-rapid technique. The front of the growing fibres has reached a position indicated by the *arrow*. Beneath the *asterisk* an invading fibre can be seen (it is shown in Fig. 54 at a higher magnification). Dendrites of radial cells located in the cortical plate penetrate the *SO*. *V*: ventricular zone, *SV*: subventricular zone, *IN*: intermediate zone, *CP*: cortical plate, *SO*: stratum opticum. *Bar* 100 μm (With permission of Springer-Verlag)

(Sect. 3). Physiological results involving a current-source-density analysis (Rager and Kreische 1978) suggest that these synapses can already be activated by optic nerve stimulation.

The discrepancy between the results obtained from light microscopic autoradiography and from Golgi-impregnation is mainly a function of the techniques used. Firstly, the number of ingrowing fibres is very small on incubation day 8 compared with day 10. Light microscopic autoradiography is able to demonstrate mass events, but it cannot resolve individual fibres. Secondly, the growing tips of fibres have invaded the SGFS to a depth of only 20–30 μm (Fig. 54, 55). Thirdly, in the grain counts of Crossland et al. (1975, Fig. 2) there is no abrupt jump from the SO to layer IX; there is a steep but continuous transition even on incubation day 9, especially at position b which is near the central tectal area. The grain counts in the upper part of layer IX seem to be several times higher than background which is consistent with our Golgi results.

The observation that central retinal fibres connect to the central tectal area can now be interpreted as follows: Having arrived at the tectum, retinal axons continue to grow over the tectal surface until they meet dendrites mature enough to receive contacts. For the first retinal fibres this is on incubation day 8 and in the central tectal

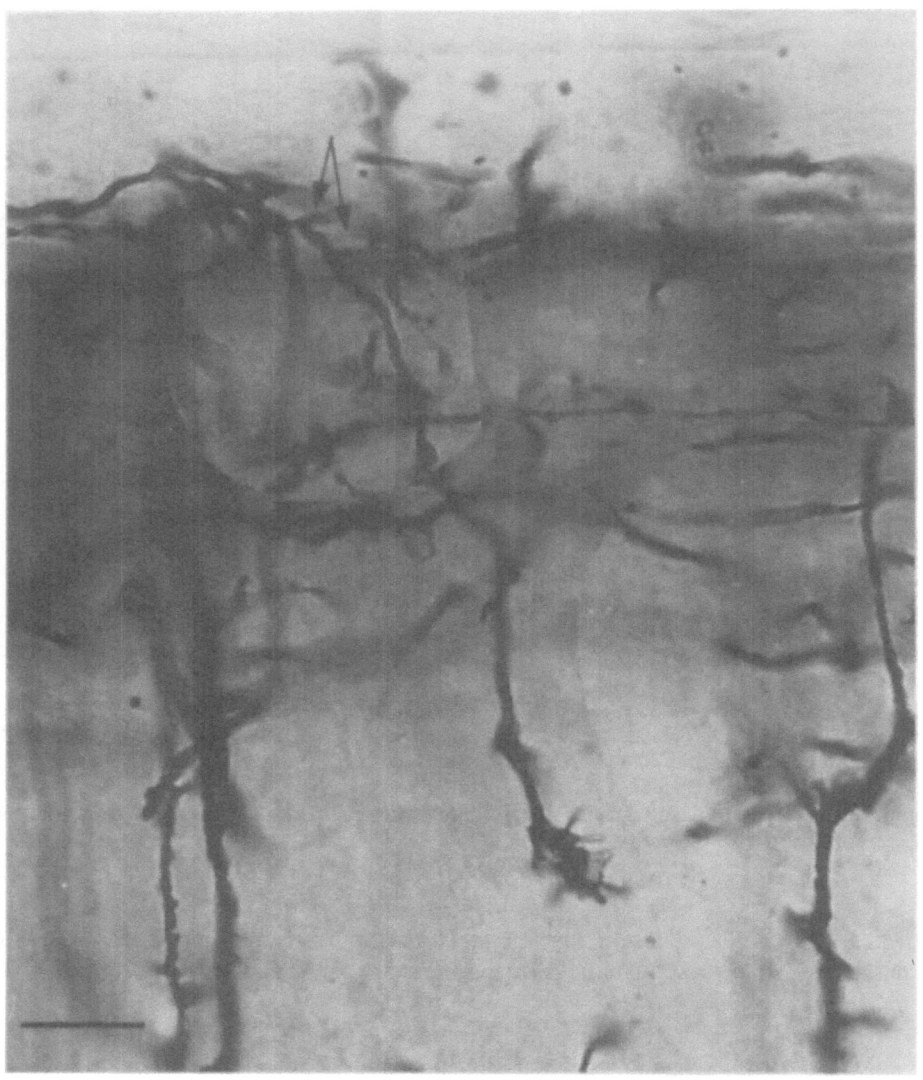

Fig. 54. Retinal fibre in the cortical plate of the central tectal area on incubation day 8. It gives off two side-branches within the SO (*arrows*). Its growth cone has already penetrated the cortical plate. *Bar* 10 μm

area. In other words, the synchrony between arrival of fibres and maturation of tectal cells could play a decisive role in determining the site at which the first connections are made. Thus, it is possible to explain why central retinal fibres connect to central tectal neurones without assuming individual cytochemical markers, which implies that the chemoaffinity hypothesis as defined above does not give the only possible explanation. However, the present discussion does not touch on the question why retinal fibres contact cells of the SGFS at all and whether they connect to them preferentially.

60

Fig. 55. Another retinal fibre invading the cortical plate in the central tectal area on incubation day 8 (*arrow*). The section is adjacent to that of Fig. 54. Abbreviations as in Fig. 53. *Bar* 50 µm

5.5 An Alternative View

We have seen that the degree of order of retinal fibres in the developing and adult visual pathway is sufficient to account for most and perhaps all of the observed retinotopic map. A rearrangement or sorting out of fibres is, therefore, not necessary. Since it is unlikely that individual markers of tectal cells are responsible for the ordered outgrowth of fibres, we have to seek other conditions which would be sufficient to establish and stabilize retinotopy and order during development. The hypothesis presented here concerns the formation of the retinotectal projection in the chicken; but it might also be applicable to other species.

Ganglion cells are generated first in the centre of the retina. Later in development generation of cells spreads to peripheral parts of the retina. Thus each ganglion cell is generated at a particular time and in a particular locality and acquires label by virtue of its coordinates in space-time.

This is underlined by the fact that ganglion cells become attached to the inner limiting membrane by transitory end-feet while they are sprouting their axons (Fig. 8, 9). If we assume that the formation of axons has approximately the same time course for all ganglion cells, there would be sufficient conditions that axons maintain their neighbourhood relations with a high probability in the retinal fibre layer as long as they grow from their site of origin straight towards the optic fissure and join already existing axons without changing their relative positions. Our Golgi-impregnated preparations show that this is, in fact, the case, but they give no clue as to the mechanism which prevents deviation of fibres. It could be that stretching of the retina generates some orientation in the substrate resulting in substrate pathways (Katz and Lasek 1979) or guidance structures (Blechschmidt 1977). The fact that retinal fibres are progressively straightened out (Goldberg and Coulombre 1972) and not only fibres but also cells show a clear radial orientation in cresylviolet stained whole mounts of the developing retina (Rager, unpublished observations) speaks in favour of this idea which has been considered in greater detail elsewhere (Goldberg 1976; Suburo et al. 1979).

The orientation of the tissue as found at the cellular level is less prominent at the entrance of the optic stalk. Correspondingly the degree of order in the fibre course is lower and the variance of the distribution function P (see Sect. 5.1) may be larger than in the other parts of the fibre pathway. Nevertheless, the fibres seem to follow the same rules as in the retina. If these rules are seen together with the special geometry of the entrance into the optic stalk, the actual transformation of incremental retinal rings into crescents in the optic stalk can be derived (Rager 1980). The secondary closure of the fibre crescents into rings and thus the re-establishment of retinotopy in the nerve may be due to another unspecific process, namely the disappearance of dorsally located neuroepithelial cells. The parallel course of retinal axons in the optic nerve, chiasm and tract could be understood if there were pre-neural guidance structures or channels (Szentágothai and Székely 1956; Goldberg 1974; Blechschmidt 1977; Katz and Lasek 1979; Silver and Robb 1979; Singer, Nordlander and Egar 1979; Silver and Sidman 1980) which may have been generated by the differential growth of the brain (Blechschmidt 1961, 1974; Blechschmidt and Gasser 1978). The principle of contact guidance (Weiss 1929, 1934, 1941, 1955) could then be applied.

Like retinal ganglion cells, tectal neurones mature first in the central area and later in peripheral regions. Individual tectal neurones are labelled in the above-defined

nse by maturing at a particular time and in a particular locality in relation to the
undle of arriving retinal fibres. The coincidence of arrival of central retinal fibres
nd maturation of dendrites of central tectal neurones would then be sufficient to
etermine the origin of the map at this specific tectal site. This hypothesis could
erhaps be tested by either delaying the arrival of the first afferent fibres or by shifting
ne tectal site where neurones mature first: the origin of the retinotectal map should
nen be different but predictable. The subsequent "appropriate" mapping onto peri-
heral tectal regions can be explained by the organization of the arriving fibre bundle
nd by the later maturation of tectal cells progressing continuously from the centre to
ne periphery. It can be shown by a special preparation that fibres lying superficially
nd laterally in the tract and on the tectal surface project to the caudal tectum,
'hereas deep fibres connect to the rostral tectum (Horder et al. 1979). Finally, the
rojecting and the receiving system can be matched in size by competition and de-
eneration as discussed in Sect. 6. These mechanisms could also be responsible for
limination of aberrant fibres and thus increase the degree of order of this map. We
ave to keep in mind, however, while discussing retinotopy and order, that the chick
etinotectal system probably is not a point-to-point, but rather a field-to-field mapping
'hich results from the convolution of the map of individual fibres with the prob-
bility density function P (see 5.1 and Fig. 39). Thus order of individual fibres has
degree which can be expressed by the variance of function P.

Since it is practically impossible for the time being to prove directly that cyto-
hemical affinities are responsible for the formation of the retinotectal map, the chemo-
ffinity hypothesis was introduced mainly by the argument that the specificity of the
etinotectal map can be explained only by assuming individual cytochemical identifi-
ation tags of retinal and tectal neurones. This argument does not hold in general,
ince specific connections do not necessarily imply cytochemically specified neurones
Gaze and Keating 1972). It does not hold in the particular case of the chicken either
ince alternative explanations are possible at the present level of morphological investi-
,ation. This view might help us to get out of the present impasse caused by the as-
umption on one single process such as chemoaffinity. A multiplicity of processes and
nechanisms could be envisaged, including molecular interactions at particular sites.

The Final Maturation of Retinal Ganglion Cells

.1 Axonal Branching and Formation of Dendrites

/hen a ganglion cell leaves the mitotic cycle, its nucleus moves towards the inner
miting membrane. The apical process loses its contact with the outer limiting mem-
rane and becomes progressively resorbed. The basal process increases in size, forms
n end-foot by which it becomes transitorily attached to the inner limiting membrane,
nd it then sprouts an axon. The perikaryon rounds out and the cell becomes unipolar.
/hile the bipolar stage of the cell is very short, the unipolar stage lasts until the axon
as reached the optic tectum. Then ubiquitous spicules and small processes appear to
row out in all directions indicating the early multipolar stage. Most of these pro-
esses disappear later on. The formation of the final dendrite seems to begin only

when axons start to branch (Sects. 3 and 5). Thus, the arrival and branching of axons may play a decisive role in the final maturation of ganglion cells. It certainly marks a "catastrophe point" (Thom 1975) in their development. This is similar to the original finding of Ramón y Cajal and Lenhossék obtained from studying the development of motoneurones in the spinal cord of chick embryos. "Dendrites appear on neuroblasts only when the axon has arrived or is about to arrive at its destination site" (Ramón y Cajal 1909, p. 611). Barron (1943, 1946) came to similar conclusions after observing the development of spinal motoneurones in sheep. The frequent occurrence of a synchrony between arrival of axons and formation of dendrites suggests that there might be a causal relationship between the two. We shall see later on (Sects. 6.2 and 6.3) that this view is supported by additional evidence.

An alternative explanation might be that "ganglion cell dendrites may differentiate in response to the formation of the amacrine cell processes" (Morest 1970, p. 63). However, this is not the case in the chick retina where no synaptic contacts can be observed in the inner plexiform layer before incubation day 13 (Hughes and La Velle 1974; our own observations). If it should be suggested that non-specific membrane adhesions (puncta adhaerentia) were responsible, no such adhesions can be found before incubation day 10 (Sheffield and Fischman 1970) which is two days after the formation of the final dendrite has begun. It can be seen from our Golgi-impregnated material that on incubation days 9 and 10 dendrites expand and ramify considerably in the central retinal area (Figs. 20, 21). A similar sequence of developmental events can be seen in the development of dendrites of tectal cells which start to differentiate before retinal fibres arrive (Fig. 53). Since the presence of afferents is not required at least for the initial phase of dendrite differentiation, we may postulate that early dendrite formation follows a metabolic gradient (Blechschmidt 1977). In general, dendrites are orientated towards the pial surface which is rich in blood vessels (e.g. retina, optic tectum, cerebral and cerebellar cortex). If cell metabolism is activated when axons have started branching (Sect. 6.2), it is possible that dendrites play an important role in balancing metabolism; structural criteria such as position and orientation indicate that dendrites could be actively involved in resorption and assimilation of molecules. The afferent input may be important for the structural development of dendrites on a finer level during these later stages.

Additional factors may contribute in working out the final shape of ganglion cell dendrites. In general, the dendrite has a single main stem which is orientated radially. After a short distance it spreads tangentially and is confined to narrow sublaminae of the inner plexiform layer (Ramón y Cajal 1893; our Golgi preparations). We have seen that the formation of the tangential dendritic arbor occurs at a time when the tangential growth of the retina is at its maximum (Rager and von Oeynhausen 1979).

6.2 Increase of Fibre Diameter as a Function of Axonal Arborization

The diameters of the largest fibre class of the optic nerve increase nearly exponentially at first. These fibres have a maximum rate of increase towards the end of the first week after hatching and reach a final size of about 3.5 μm two months after hatching (Fig. 35). How can this phenomenon be interpreted? Is the fibre diameter increase directly related to an increase of fibre length, as has been suggested by several authors? This suggestion cannot be supported here for the following reasons: Firstly, in the arge peripheral nerves of adult animals, the whole range of diameters of unmyelinated and myelinated fibres is present, although all fibres have similar lengths. Secondly, during ontogenesis the increase in the length of the optic fibre pathway connecting the retina and the optic tectum attains its maximum value on incubation day 11 (Fig. 54), when the increase in fibre diameters is very small (Sect. 6.3).

We may be able to solve this problem if we consider the fact that axons are transport channels between two systems, the retina and the optic tectum. We assume here that the transport capacity of each individual channel is correlated with the peripheral volume supplied. The number of axonal terminal branches can be used as a measure of this peripheral volume. We are now able to develop a model which could explain the observed increase in axon diameters.

We know (Goldberg 1974; our own observations) that optic nerve fibres in general do not form collaterals before they have reached their termination site. Thereafter these axons make a more or less right-angled turn, invade the optic tectum, and start to ramify (Sect. 5.4).

We assume that the axon diameter d is a function of the axonal flow J which is the quantity of material transported per unit time. This flow comprises both antero-grade and retrograde transport. That is,

$$d = f(J) \tag{9a}$$

where J is measured in moles per unit time. In addition, we assume that the flow J depends linearly on the number of terminal axon branches N

$$J = \alpha \times N + \beta \tag{9b}$$

where α and β are constants. Finally it is assumed that the flow density j (flow per unit area A) remains constant throughout development:

$$j = J/A = \text{constant} \tag{9c}$$

and so the flow J is proportional to the cross-sectional area and hence to the square of the axon diameter. From Eqs. (9a), (9b), and (9c) we obtain

$$d = a \times N + b \tag{9}$$

The values of the constants a and b can be calculated from initial and final values of fibre diameters (d_0 is the axon diameter at $t = t_0$ which is taken to be incubation day 8; d_∞ is the diameter at $t = t_\infty$ which is taken to be 100 days after hatching) and the corresponding numbers of terminal axon branches are N_0 and N_∞. The minimal and maximal diameters are known from our measurements ($d_0 = 0.25\ \mu m$; $d_\infty = 2.36\ \mu m$ and $3.375\ \mu m$ for axon and fibre diameters respectively). Since only a few axons have collaterals before they arrive at the optic tectum and start branching on incubation day 8, we put $N_0 = 2$. The maximal number of terminal branches can be estimated from drawings by Ramón y Cajal (1911) and La Vail and Cowan (1971a). We obtain a value for N_∞ which is between 50 and 80. We want to emphasize that the number of terminal branches is used for the model, not the number of synapses.

It is clear from Golgi preparations that branching of axons is mainly dichotomous. We assume that each terminal branch grows at the same rate and that at any instant each newly formed terminal branch has the same probability of forming a new bifurcation. Since we have to deal with a large number of axons, we avoid using a step function needed for single axons and proceed directly to the corresponding continuous function which is written as follows:

$$\frac{dN}{dt} = k \times N_a \tag{10}$$

where N_a is the number of terminal branches present and able to bifurcate and k is a constant of dimension $1/t$.

As time goes by more and more terminal branches will be prevented from further ramification; this could happen, e.g. by synapse formation and by spatial or metabolic limitations, so that only free endings are available for further ramification. Thus,

$$N_a = N - S \tag{11}$$

where S is the number of blocked telodendra. As a first approximation we assume that the probability P that a particular branch will be blocked is proportional to N, thus $P = B \times N$, where B is a constant. The number of blocked telodendra S is then given by

$$S = P \times N = B \times N^2 \tag{12}$$

Combining Eqs. (10) and (11) we obtain

$$\frac{dN}{dt} = k \times N \times (1 - BN) \tag{13}$$

We can calculate the constants k and B from the boundary conditions already discussed. Thus, k is $(N_\infty/N_0) - 1$ and B is $1/N_\infty$. The solution of Eq. (13) is then

$$N = \frac{N_\infty}{1 + (N_\infty/N_0 - 1) \times \exp(-k \times t)} \tag{14}$$

From Eq. (9) the solution for the diameter d is:

$$d = \left| \sqrt{\frac{d_\infty{}^2 - d_0{}^2}{N_\infty - N_0} \times (N - N_0) + d_0{}^2} \right. \tag{15}$$

The only unknown parameter of Eq. (15) is k. Its value was found by fitting this function to the measured fibre diameters. If the optimization is done with $N_\infty = 80$, k becomes 0.162 [day^{-1}], 2.83×10^{-3}. The 95% confidence interval is 6.025×10^{-3} (computed according to Zurmühl 1965), the coefficient of determination $r = 0.998$. Finally we know that the ratio $\rho = d/D$ (D is the outer diameter of fibres) is in the range 0.6–0.8 during ontogenesis (Rager 1976a). Thus, the mean value of $1/\rho$ is taken to be 1.43. From this we can calculate the mean outer diameter of myelinated fibres using Eq. (16)

$$D = \frac{1}{\rho} \times d \tag{16}$$

With this model we obtain the graph in Fig. 35. On incubation days 16 to 18 myelination begins and Eq. (16) becomes valid, hence the jump seen in the graph.

While developing the model we tried out various relationships between d and N but found that the square-root function (Eq. 9) gave the best fit. Thus, the assumption expressed in Eq. (9c) is supported: that the axonal flow is proportional to the axonal cross-sectional area. This would imply that the total flow along the axon is constant for a given cross-sectional area, i.e. transverse inflow and outflow which might depend on metabolism and membrane transport do not substantially change the total flow. Finer details are neglected, since they are unknown.

6.3 Competition and Degeneration

Since dendrites develop only when axons have arrived at the optic tectum and since in the model the increase in fibre diameters depends on the increase of the axonal periphery, we suggest that the final maturation, and hence also the survival of retinal ganglion cells, depends on the axonal branching process. Thus, if there is no termination site available when a fibre reaches its final termination area, the cell will die. This prediction can be tested if there is an abundance of ganglion cell axons over termination sites in the tectum during development. The matching of the two systems, retina and optic tectum, would then be accompanied by death of supernumerary ganglion cells (the term "systems-matching" is used in analogy to the definition introduced by Gaze and Keating 1972). With this in mind, we have checked whether degeneration of ganglion cells occurs during the formation of retinotectal connections and if so, whether it is correlated with axonal branching. This problem has been considered in the context of the more widespread phenomenon of cell death seen during ontogenesis which has been reviewed elsewhere (Rager and Rager 1978).

6.3.1 Degeneration of Ganglion Cell Perikarya

6.3.1.1 First Appearance of Degenerating Cells

We have examined the ganglion cell layer of the retina with the electron microscope during the whole period of embryonic development. No degenerating ganglion cells were found before incubation day 8. On incubation day 8, a few degenerating cells could be seen for the first time in the central retinal area. Figure 56 shows an example of a degenerating cell lying next to apparently normal ganglion cells. The debris is already engulfed by glial cell processes. By incubation day 9, there were a considerable number of degenerating perikarya. Two examples of degenerating cells from incubation day 10 are shown in Fig. 57 and 58. The degenerating cell in Fig. 57 is at a relatively early stage of degeneration; it is located near the optic fibre layer. It is already split up into several fragments and contains a large osmiophilic body as seen in other degenerating cells (Raisman 1973; Chu-Wang and Oppenheim 1978) which probably consists of condensed chromatin. These fragments are surrounded by glial cells. The two cells seen in Fig. 58 are already at an advanced stage of degeneration. Only amorphous cellular debris can be recognized. We know from our Golgi studies that branching of ganglion cell axons can be observed in the tectum already on incubation

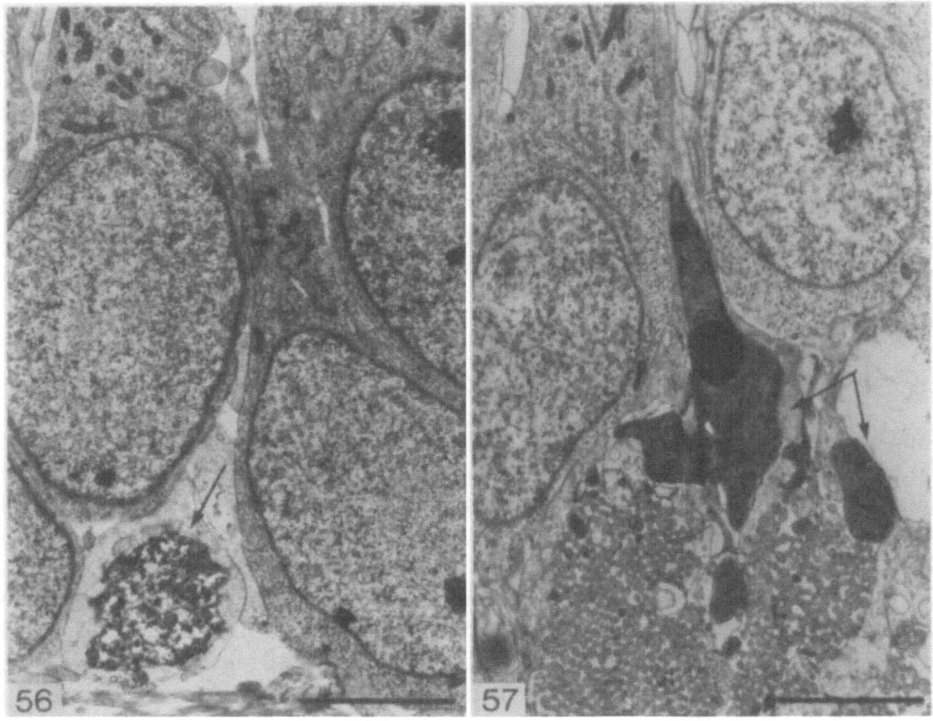

Fig. 56. A degenerating cell in the central retinal area on incubation day 8. Cellular debris is engulfed by glial processes (*arrow*). *Bar* 3 μm

Fig. 57. Fragments of a dying cell at an early stage of degeneration (*arrows*). Central retinal area, incubation day 10. *Bar* 4 μm

day 8 (Rager and von Oeynhausen 1979; and also 5.4). Since this is when degenerating perikarya can be seen for the first time, we may conclude that ganglion cells do not degenerate before their axons have arrived at the optic tectum and have started branching. A new stage in the maturation of ganglion cells appears to begin with the axonal branching process. We propose, therefore, that ganglion cells become competent either to continue maturation or to degenerate only after they have arrived at their termination sites and we call these cells "competent" cells. This view is supported by the fact that synchronously with synapse formation a well-organized rough endoplasmic reticulum is formed (Pilar and Landmesser 1976) which indicates that important changes in cell metabolism result from events in the axonal periphery.

6.3.1.2 Topography of Degenerating Cells

Electron microscopic examination provides information about the size, shape, and appearance of degenerating cells. This permits localization of these cells with the light microscope. To localize degenerating cells at various stages of development, whole eyes were cut at a defined section plane (see Sect. 2), the retinae were examined under an oil-immersion objective lens, and the position of each degenerating cell was record-

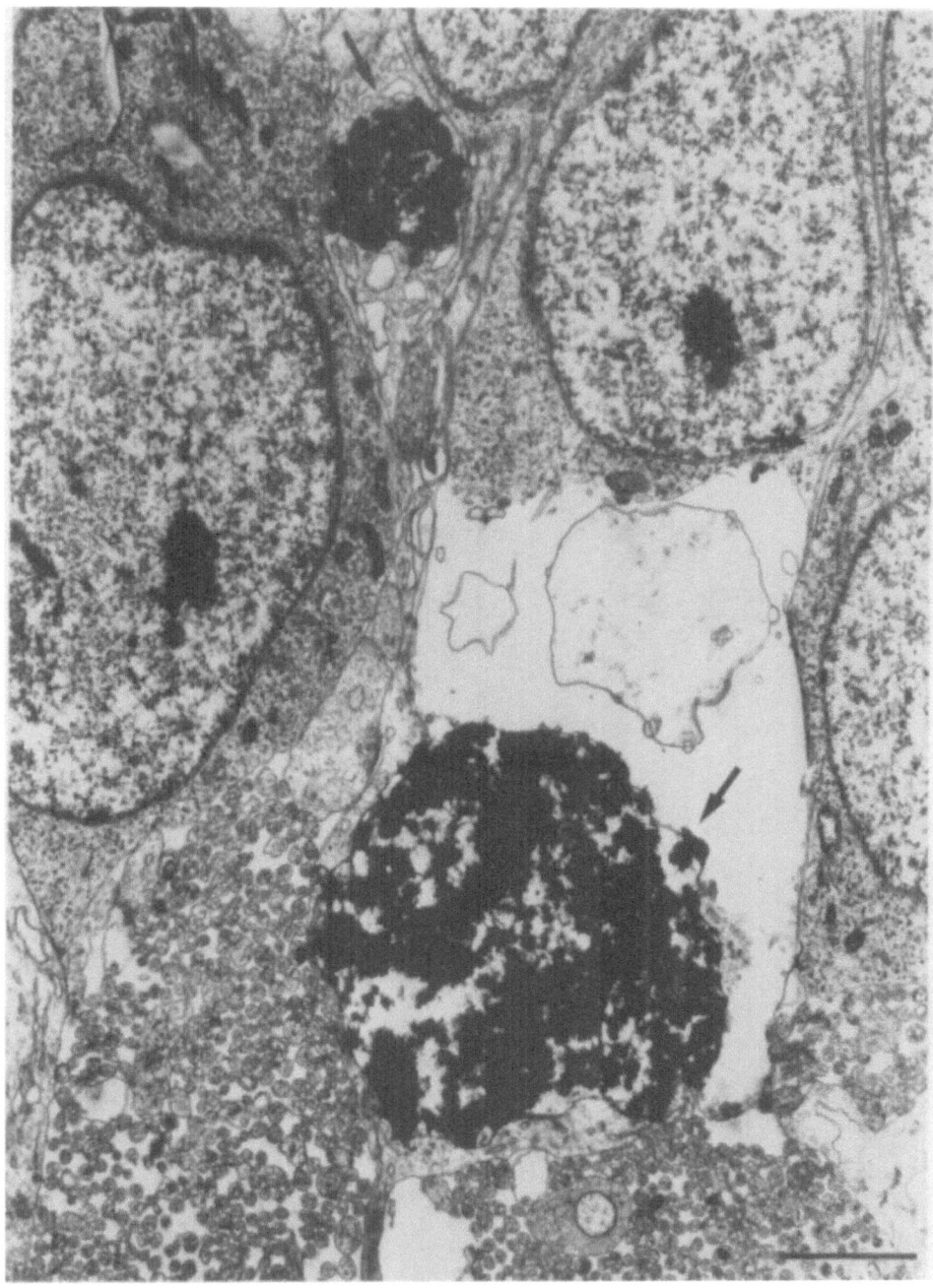

Fig. 58. Two degenerating cells at an advanced stage of cell death (*arrows*). Central retinal area, incubation day 10. *Bar* 2 μm

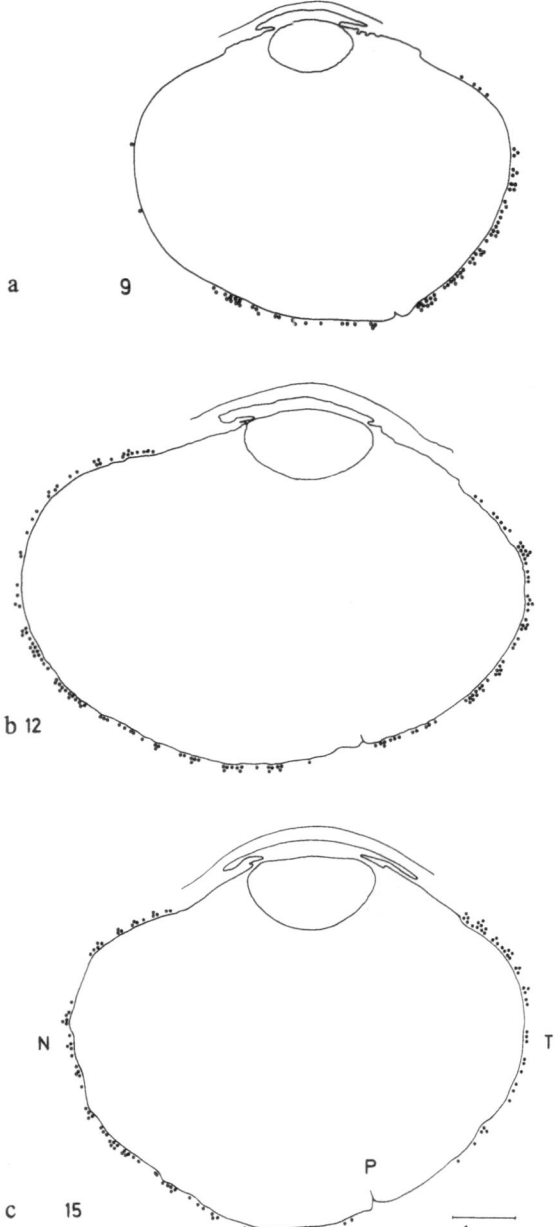

Fig. 59a–c. Distribution of degenerating cells in the retina on incubation days 9 (a), 12 (b) and 15 (c). The eyes were cut through the centre of the lens and perpendicular to the papilla (*P*). Each degenerating cell found (oil immersion) is localized with a *dot*. *N*: nasal, *T*: temporal

a   9

b 12

N

T

P

c   15

1mm

ed on a plotter connected to the stage control with linear potentiometers. Three developmental stages were selected for Fig. 59. During the main period of cell death degenerating cells are distributed over a large part of the retina, but there are more in some zones than in others. In the early phase (incubation day 9) they are concentrated in the region around the papilla (Fig. 59a). Three days later, when degeneration has attained its highest rate, they are distributed over the whole retina, but are slightly less concentrated in the central area (Fig. 59b). On incubation day 15, when degeneration begins to fade away they are found mainly in more peripheral zones (Fig. 59c). Thus, the wave of ganglion cell generation (Kahn 1974; see Sect. 5) is followed by a similar wave of degeneration which spreads from the centre to the periphery of the retina. This wave of degeneration is closely correlated with the mode of ingrowth of retinal fibres into the optic tectum which is consistent with the idea (put forward in Sect. 5.3.1.1) that ganglion cells do not degenerate before their fibres have arrived at their termination area or before they have become competent in the above defined sense.

5.3.2 Degeneration of Ganglion Cell Axons

5.3.2.1 The Total Number of Fibres

To get more insight into the processes underlying degeneration, we need quantitative data on the number of degenerating cells and the time course of degeneration. Counting the number of neurones in the ganglion cell layer would entail severe problems. Firstly, quite a number of displaced amacrine cells are located in the ganglion cell layer (Ramón y Cajal 1911; Binggeli and Paule 1969); it would be extremely difficult to evaluate the percentage at the various developmental stages. Secondly, it is difficult to distinguish between glia and nerve cell nuclei. Thirdly, cell counts are usually done in terms of nucleolus counts. Since the number of nucleoli per nerve cell nucleus changes considerably with developmental age (Zilles et al. 1976), correction factors have to be introduced which can be an additional source of error (Cowan 1973; Zilles 1978).

We have sufficient evidence from Golgi-impregnated whole mounts and from tangential sections of the Golgi-impregnated optic nerve that ganglion cell axons have practically no collaterals in the retina or in the optic nerve. This view is supported by other observations based on silver impregnations (Goldberg and Coulombre 1972; Goldberg 1974). Thus, we should be able to estimate the number of ganglion cells by counting the number of fibres in the optic nerve. In addition, since the sprouting of an axon is one of the first events in the development of retinal ganglion cells, total fibre counts reflect not only the degeneration, but also the generation of these cells. We therefore have estimated the total number of fibres in optic nerve cross-sections from incubation day 5 until 100 days after hatching using the electron microscope (see Sect. 2).

The youngest animal in which the total number of optic nerve fibres was determined was a 5-day-old embryo (stage 26). At this age, the first fibres have crossed the chiasm, but have not yet reached the optic tectum. The cross-sectional area of the whole optic nerve is so small that it could be covered with 40 micrographs enlarged to a total magnification of 8000. The entire cross-section was reconstructed in a photomontage and all fibres were counted. The total number of fibres was 4400, while the cross-sectional area of the ultrathin section was 13 800 (μm^2). During the following days (Fig. 60, Table 4), the total number of fibres increases nearly exponentially until incubation day 8. From then the increase is slowed down. On incubation days 10 and 11 a maximum number of 4 million fibres is reached. Only one day later the total number of axons has started to decrease. The reduction is rapid, reaching a maximum rate on incubation day 13; it is then more gradual. By incubation day 18, the total number of fibres is reduced to a value which then remains constant at about 2.4

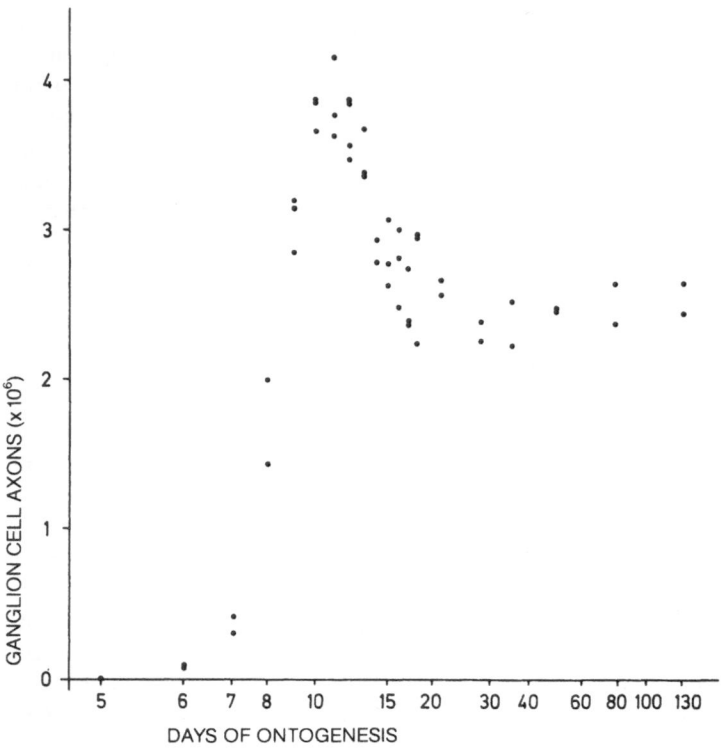

Fig. 60. Total numbers of optic nerve fibres plotted on a logarithmic time scale. Each of the 50 *dots* represents the nerve of one animal

Table 4. The increase of the number of nerve fibres as a function of incubation time

Days of onto-genesis	Animal No.	Fibres counted No.	Fibre density [N/1000 μm^2]	Area of the ultrathin section [μm^2]	Fibres, nerve Total no.
5	1	4 400	320	13 800	4 400
6	1	3 178	1 598	54 873	87 658
	2	3 018	3 080	35 475	109 263
7	1	8 349	3 960	81 996	325 000
	2	5 068	5 880	73 703	433 374
8	1	18 377	12 000	165 982	1 991 247
	2	7 361	5 640	115 287	880 000
	3	12 676	9 270	155 559	1 442 032
9	1	17 586	17 980	174 615	3 139 578
	2	25 722	16 510	172 352	2 845 532
	3	16 401	16 040	193 601	3 105 360
10	1	12 320	12 830	286 188	3 672 000
	2	10 164	10 660	361 789	3 856 671
	3	17 812	16 480	235 507	3 881 155

Table 4 (continued)

Days of onto-genesis	Animal No.	Fibres counted No.	Fibre density [N/1000 μm^2]	Area of the ultrathin section [μm^2]	Fibres, nerve Total no.
11	1	41 643	11 340	320 062	3 629 503
	2	11 727	12 240	309 176	3 784 314
	3	12 120	11 650	357 602	4 166 063 .
12	1	10 342	10 650	335 175	3 569 614
	2	13 097	13 290	261 818	3 479 561
	3	8 964	8 930	430 880	3 847 758
	4	8 999	8 600	449 156	3 862 742
13	1	9 196	8 520	398 299	3 393 507
	2	8 364	8 000	459 926	3 679 408
	3	11 010	8 360	403 072	3 369 682
14	1	14 558	7 270	382 936	2 783 355
	2	8 455	8 190	357 793	2 930 325
15	1	12 624	5 750	455 644	2 620 653
	2	6 438	6 330	485 038	3 070 290
	3	6 053	5 600	494 498	2 769 188
16	1	18 086	4 200	590 304	2 479 277
	2	7 789	4 850	619 348	3 003 838
	3	8 512	6 030	466 509	2 813 049
17	1	12 351	4 950	553 237	2 739 000
	2	6 230	3 780	633 419	2 566 662
	3	8 358	5 150	460 125	2 369 644
18	1	22 967	5 160	435 185	2 244 037
	2	9 153	5 040	585 933	2 953 102
	3	4 767	4 290	691 542	2 966 715
21	1	7 395	4 260	623 815	2 658 578
	2	5 467	3 140	815 724	2 561 373
28	1	13 914	2 600	913 784	2 375 838
	2	9 462	1 970	1 140 620	2 247 021
35	1	8 366	1 930	1 153 110	2 221 339
	2	10 136	2 410	1 043 750	2 524 039
50	1	7 884	1 540	1 600 410	2 465 000
	2	7 776	1 710	1 440 000	2 458 220
78	1	8 640	1 150	2 292 300	2 627 000
	2	4 207	680	3 481 056	2 367 118
119	1	9 688	980	2 683 150	2 629 487
	2	4 543	720	3 363 693	2 430 268

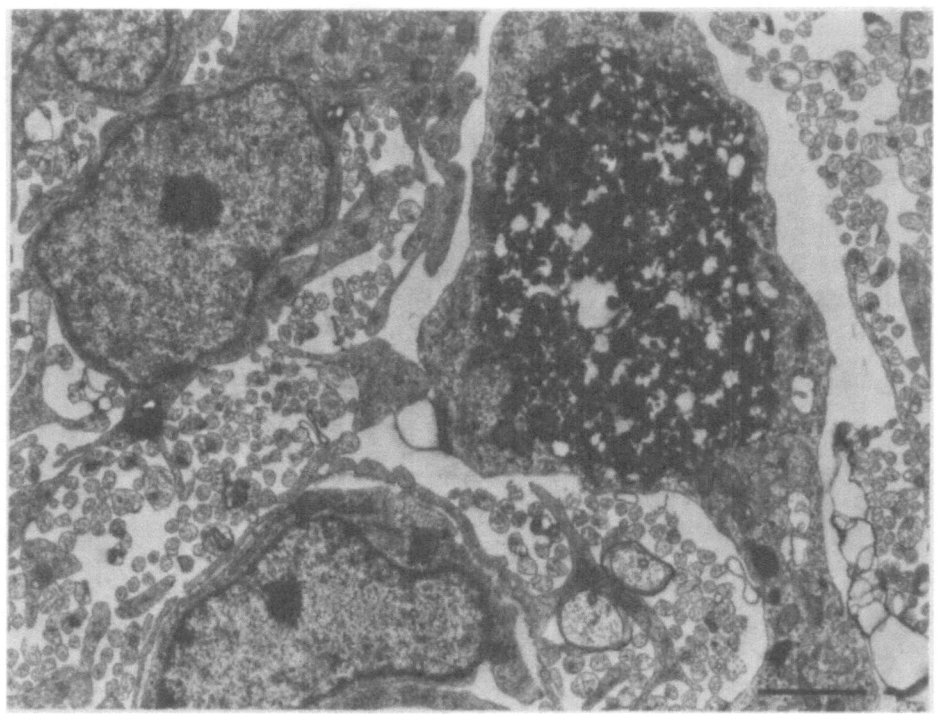

Fig. 61. The optic nerve on incubation day 16. A few phagocytes can be seen to contain dark circular inclusions which are probably debris of degenerating axons. *Bar* 1.5 μm

million. This is about 40% less than the maximum. Both the formation and the degeneration of retinal ganglion cells end before the visual system begins to transmit light signals at incubation day 18 (Rager 1977, 1979).

Cell death in the retina could be detected three days earlier than degeneration of axons in the optic nerve for two main reasons. Firstly, it is difficult to distinguish degenerating fibres from fixation artefacts. Therefore, axons were counted as long as they could be identified. Secondly, generation and degeneration of ganglion cells overlap for a while so that degeneration cannot be recognized from total numbers alone before incubation day 12.

In addition to degenerating axons, we occasionally found cells full of ring-like inclusions and dark elements in the optic nerve. They occur mostly at later stages of development. It is probable that they are phagocytes which are removing debris of degenerating axons (Fig. 61).

6.3.2.2 Fibre Density and Cross-sectional Area of the Nerve

The total number of fibres can be obtained by multiplying fibre density by the respective cross-sectional area. Fibre densities and cross-sectional areas taken independently of each other give relative rather than absolute values if, as here, sections are not corrected for shrinkage and compression (see Sect. 2). Nevertheless, important results can be obtained from these relative values. The fibre density of the nerve (Fig. 62,

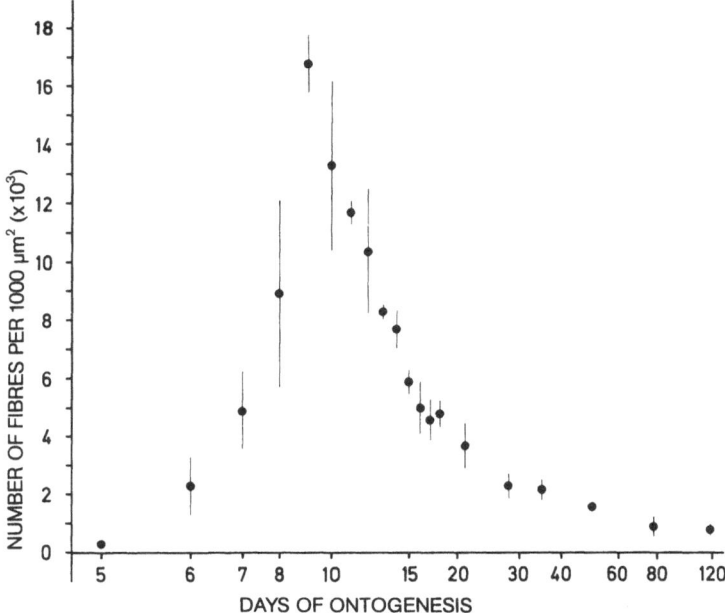

Fig. 62. Fibre density values (number of fibres per 1000 μm^2 together with their standard deviations) are plotted on a logarithmic time scale

Table 4) is 320 fibres per 1000 μm^2 on incubation day 5. It increases very rapidly and reaches its maximum value of 16 800 fibres/1000 μm^2 by incubation day 9. Thereafter it decreases rapidly at first. From about incubation day 18 onwards the decay slows down and approaches the final value of about 1000 fibres per 1000 μm^2 asymptotically.

The time course of fibre density values is quite different from the time course of total fibre counts; thus, fibre density may be influenced by factors other than those which determine the total number of fibres.

Firstly, the percentage of area covered by neuroepithelium is reduced rapidly and the percentage of growth cones in the fibre population decreases. This contributes to the rapid ascending phase in fibre density. Secondly, as shown in Sect. 3, on incubation days 9 to 12 small unmyelinated fibres having roughly the same diameter are packed together in large bundles; only a few glial processes separate these bundles from one another. Only when axon diameters increase, do the glial cells begin to sprout. Thirdly, the tissue compartment consisting of blood vessels increases rapidly concomitantly with the growth of glial cells. The increase of axon diameters, the myelination of axons, the expansion of glial cells and the increasing number of blood vessels contribute to the reduction of fibre density which attains its final value between two to three months after hatching.

These developmental events are reflected also in the increase of the cross-sectional area of the optic nerve (Fig. 63, Table 4). The values, plotted here on a double logarithmic scale, can be fitted by adding two sigmoidal curves arising from two growth functions. The first curve reflects the increase in the total number of fibres. If the total number of fibres N is multiplied by their mean cross-sectional area A_0, which is proportional to the compartment occupied by these fibres, we obtain the

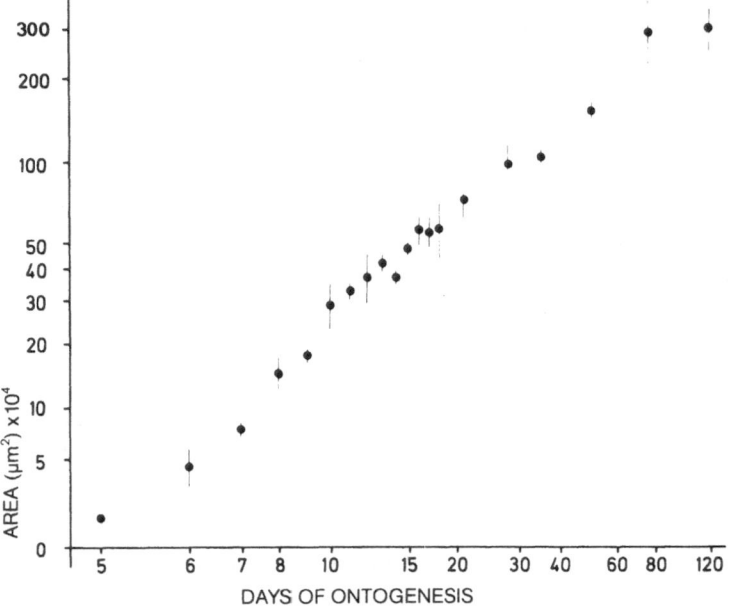

Fig. 63. The cross-sectional area of optic nerves measured from ultrathin sections is plotted against time of development on a double logarithmic scale

fractional area F_1 as

$$F_1 = a \times N \times A_0 \tag{17}$$

where a is a constant.

The second sigmoidal curve mainly reflects the increase of fibre diameters. Let N_S be the number of surviving cells. At any given time only the newly formed fraction of N_S increases in diameter. This leads to the following convolution integral

$$F_2 = N_S \times A = \int_{t_0}^{t} \frac{dN_S}{dt} \times [A(t - \tau) - A_0] \times d\tau \tag{18}$$

where A refers to the cross-sectional area of developing fibres which is proportional to the square of fibre diameters as derived in Eqs. (15) and (16). Thus, for the total cross-sectional area of the nerve we obtain

$$F = F_1 + F_2 \tag{19}$$

The actual sigmoidal curves deviate, however, from the pure growth functions presumably because of additional factors such as decrease of neuroepithelium, gliogenesis, and formation of blood vessels. This deviation was found by fitting the function defined by Eq. (19) to the measured values. The theoretical curves agree with the measured values in that the cross-sectional area of the nerve decreases slightly between incubation days 13 and 15 because the decrease in the total number of fibres temporarily outweighs the increase of fibre diameters (Fig. 63, Table 4).

6.3.3 Interpretation With a Mathematical Model

The quantitative data presented in the previous section provide a reliable basis for testing the hypothesis, already suggested by the fibre-diameter model (Eq. (15)), that the final maturation and survival of ganglion cells depends on the axonal branching process. Since there is a relationship between branching and formation of synaptic contacts, this would be equivalent to saying that fibres tend to establish a minimum number of synaptic contacts. If this subsistence minimum is not reached, the cell will die. By this mechanism the two systems, retina and optic tectum, are matched in size. The hypothesis which involves systems-matching by degeneration on the basis of competition is expressed in a mathematical model. The validity of the model can be tested by measuring how well it fits the present quantitative data.

The model consists of a system of differential equations which represent the generation of retinal ganglion cells and termination sites in the optic tectum. Since degeneration does not occur before ganglion cell axons have arrived at their termination site in the tectum and started branching, a new stage in the development of ganglion cells is defined by this arrival time. After the generation of ganglion cells, time is required for fibres to travel from the retina to the optic tectum and only then can the cell become competent. The equations for the formation of competent ganglion cells and of tectal termination sites are related by the principle of competition which states that at any given time cells which are in excess of the number of termination sites have to die. The mechanism underlying the formation of retinal ganglion cells is cell division. Therefore, the differential equation for the rate of change in the number of ganglion cell axons (Eq. 20) is formally equivalent to that used earlier for the fibre-diameter function (Eqs. (10)–(14)). The more mitotic cells there are present, the more daughter cells will be generated.

At the same time the probability increases that cells leave the mitotic cycle and differentiate as ganglion cells. These cells are no longer available for further division, thus it follows that

$$\frac{dN}{dt} = k_1 \times N \times (1 - \frac{N}{N_\infty}), \tag{20}$$

where N is the number of ganglion cell axons at the site of observation in the optic nerve, N_∞ is the final number, k_1 is a constant of dimension $1/t$. This has as solution

$$N = \frac{N_\infty}{1 + (\frac{N_\infty}{N_0} - 1) \times \exp(-k_1 \times t)} \tag{20a}$$

The formation of competent cells is delayed by the time taken by fibres to grow from the site of observation to their termination area in the tectum. The elongation of the fibre pathway during ontogenesis can be understood in terms of the volume growth of the brain

$$H = a \times V^{1/3} \tag{21a}$$

where H is the actual length l of the fibre path, a is a constant and V is brain volume. To a first approximation we assume that the volume growth of the brain is proportional to the increase in the number of cells; the more mitotic cells there are, the greater the increase of cell number will be. With time, however, more and more cells leave the mitotic cycle and mature so that the formation of new cells will gradually be reduced. This process can be expressed in terms of volume thus:

$$\frac{dV}{dt} = k_2 \times V \times (1 - \frac{V}{V_\infty}),$$ (21b)

where k_2 is a constant of dimension $1/t$ and V_∞ is the brain volume of the adult animal. From Eqs. (21a) and (21b) the elongation of the fibre pathway can be derived:

$$\frac{dH}{dt} = \frac{1}{3}k_2 \times H \times (1 - (\frac{H}{H_\infty})^3).$$ (21)

This equation has the solution

$$H = \frac{H_\infty}{\{1 + ((\frac{H_\infty}{H_0})^3 - 1) \times \exp(-k_2 \cdot t)\}^{1/3}}$$ (21c)

where H_0 and H_∞ are the retinotectal distances in the chicken at time t_0 and t_∞, respectively. Having done the Gauss-Newton iteration we obtain a coefficient of determination $r = 0.995$. The parameter values are:

$$H_0 = 6.64 \text{ mm } (t_0 = 8), \ H_\infty = 12.65 \text{ mm}, \ k_2 = 0.197 \text{ [day}^{-1}\text{]}.$$

The experimental data are given in Fig. 64 together with the pathway function and its normalized derivative.

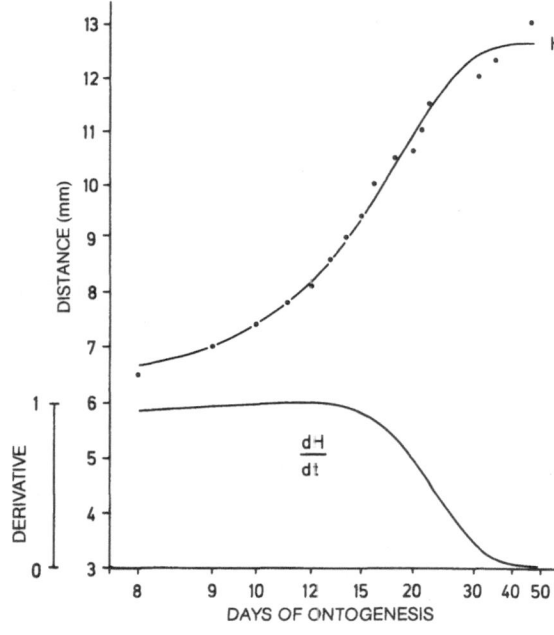

Fig. 64. The distance between the optic nerve papilla and the lateral prominence of the optic tectum plotted on a logarithmic time scale. The graph shows the results expected from the fibre-pathway function (Eq. 23). The lower curve shows the normalized derivative with respect to time of the fibre path length H. (With permission of Springer-Verlag)

78

To get the actual time required by fibres to reach their termination site, we have to know how fast ganglion cell axons grow. The rate of axonal elongation v, determined mainly from tissue culture studies, has a maximum of 1.2 mm/day (Hughes 1953). In situ, however, higher values have been measured ranging from 1.5 to 3 mm/day (Clarke and Cowan 1976) and 1.9 to 2.4 mm/day (Lund and Bunt 1976). Similar values have been obtained in regeneration experiments (Ramón y Cajal 1913). We have chosen an intermediate value of 2 mm/day.

On dividing the rate of extension of the pathway (dH/dt) by the growth velocity v of axons, we get the retardation factor at each individual stage. The rate of extension of the pathway is always smaller than the growth velocity of axons; in other words $dH/dt \times 1/v < 1$. Thus, we have to replace then the time variable t by $(t - H/v)$ in Eq. (20) in order to get the actual number of competent cells (N_C). The differential equation for competent cells can then be written thus

$$\frac{dN_C}{dt} = N_C \times (1 - \frac{N_C}{N_\infty}) \times k_1 \times (1 - \frac{dH}{dt} \times \frac{1}{v}) \tag{22}$$

the solution of which is

$$N_C = \frac{N_\infty}{1 + ((\frac{N_\infty}{N_0} - 1) \times \exp(-k_1 \times (t - H \cdot \frac{1}{v})))} \tag{22a}$$

The initial value N_0 is that seen when the first fibres have reached their termination site on the optic tectum, that is, when $(v \times t - H) = 0$.

We could not find any data giving the number of tectal cells in the layers where ganglion cell axons terminate. We therefore had to introduce a new quantity, the "termination domain" which is at present unknown but can be subjected to further investigation. A termination domain S of a retinal fibre in the tectum can be defined as the average amount of space occupied by an axonal tree forming synaptic contacts. This domain is assumed to be proportional to an unknown, but constant, number of tectal cells. Therefore, the increase of the number of termination domains can be expressed in terms of cell proliferation by analogy to the generation of retinal ganglion cells. The time required by tectal cells to migrate from the ventricle to their final position and to produce a dendrite can be said to be approximately the same in each tectal region. It will be included in the time constant $1/k_3$ of dimension t. It has to be emphasized, however, that proliferation progresses from the rostro-ventro-lateral portion of the tectum to the caudo-dorso-medial pole (LaVail and Cowan 1971b) as does the related progression of maturation of visual tectal layers and thus termination domains. In formal analogy to Eq. (20), we can express the generation of termination domains S in the following differential equation

$$\frac{dS}{dt} = k_3 \times S \times (1 - \frac{S}{S_\infty}) \tag{23}$$

which has the solution

$$S = \frac{S_\infty}{1 + (\frac{S_\infty}{S_0} - 1) \times \exp(-k_3 \times t)} \tag{23a}$$

The final number S_∞ equals the final number of fibres in the adult chicken, which is 2.4×10^6.

The central idea in our model can be expressed as follows. All retinal fibres which are in excess of those needed for the available tectal termination domains at any given time degenerate with a time constant $1/\alpha$ of dimension t. The excess is the number of competent fibres N_C minus the number of fibres already degenerated N_D compared with the number of available tectal domains S. The differential equation then becomes:

$$\frac{dN_D}{dt} = \alpha \times (N_C - N_D - S) \tag{24}$$

with the solution

$$N_D = \exp(-\alpha t) \times [N_{D_0} + \alpha \times \int_{t_0}^{t} (N_C - S) \times \exp(\alpha t)\, d\tau] \tag{25}$$

N_{D_0} is the number of degenerating fibres at time t_0. It is set equal to zero.

The number of fibres (Np) present at any given time equals the number of generated cells minus the number of degenerated cells, so that

$$Np = N - N_D \tag{26}$$

This equation determines the shape of the curve in Fig. 65.

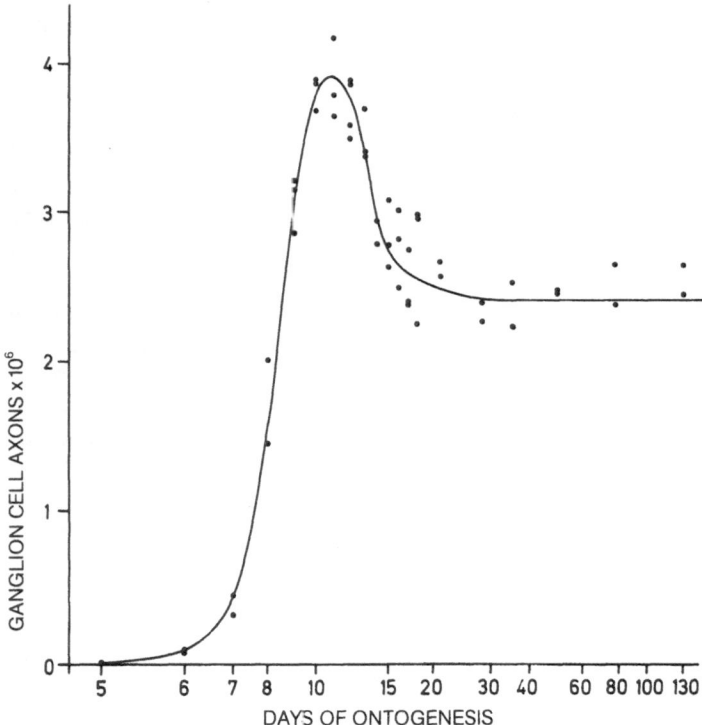

Fig. 65. The total number of ganglion cell axons plotted on a logarithmic time scale. The points represent the measured values. The graph was obtained from the model function (Eq. 26). (With permission of Springer-Verlag)

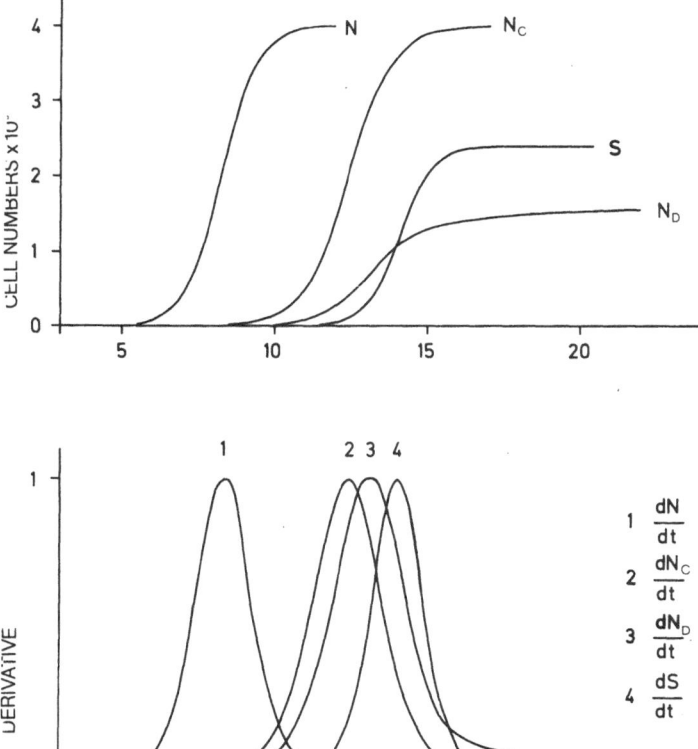

Fig. 66a. The numbers of generated ganglion cells (N), of competent cells (NC), of degenerating cells (N_D), and of tectal termination sites (S) are plotted against time, b) the normalized derivatives of these functions with respect to time

The optimal parameter values obtained by the Gauss-Newton method, their standard errors (SE) and the coefficient of determination r are as follows: $N_0 = 5.13$, SE = 7.23, $N_\infty = 4\,026\,140$, SE = 102 006, $k_1 = 1.64$ (day^{-1}), SE = 0.17, $k_3 = 1.86$ (day^{-1}), SE = 0.085, $\alpha = 0.247$ (day^{-1}), SE = 0.096, $r = 0.998$.

Since S_0 is linearly dependent on k_3, the starting value ($S_0 = 1 \times 10^{-5}$ for $t_0 = 1$) found with the lattice point procedure was kept constant.

There is no doubt that α is the most interesting parameter. The time constant $\tau = 1/\alpha$ for the degeneration of a ganglion cell axon is approximately 4 days; this takes into account the competition for a suitable termination domain, the degeneration process itself as well as the removal of cell debris.

A number of quantitative results can be obtained from the model (Fig. 66). On incubation day 4 approximately 3700 fibres should be present in the optic nerve at the site of our cross-cut. The rate of increase in the number of fibres reaches its maximal value in the course of incubation day 8 and the total number reaches saturation (99% of the maximal value) by incubation day 11.

Meanwhile degeneration has begun. On incubation day 10, the number of degenerating fibres is 2.5×10^4. On day 13, the maximal rate of degeneration is reached

81

$(4 \times 10^5$ fibres/day), and one week after hatching the process of degeneration has come to an end.

The rate of extension of the fibre pathway dH/dt reaches its maximum in the course of incubation day 11. Extension goes on for quite a long period after hatching. In spite of this long duration of brain growth, all fibres (N_C) have reached their termination site on incubation day 16.

The formation of termination domains in the optic tectum may also be described. Starting with approximately 1200 domains on incubation day 10, the final value is reached by the end of incubation day 16. The highest rate of formation is found in the course of day 14.

The maximum increase in the number of competent fibres and the maximum increase in the number of termination domains are clearly separated on the time axis (incubation days 12.5 and 14, respectively). Thus, the ratio between the number of arriving fibres and the number of termination sites varies continuously (Fig. 66). The central fibres, which arrive first, encounter a less favourable situation than the peripheral ones arriving later. This predicts that degeneration is heavy to begin with and has already attained its maximal rate by the end of incubation day 12 which is in fact the case.

6.3.4 Competition, Synaptogenesis and Selective Stabilization

The validity and the implications of this model are discussed elsewhere (Rager 1978). We have to keep in mind that the mechanisms underlying competition are still unknown. Therefore, we use the term competition in an operationally defined sense. We say that a minimum number of target sites is required for the survival of a neurone. Fibres compete for this subsistence minimum. Those fibres which do not succeed have to die.

Degeneration starts earlier than synaptogenesis according to both morphological and electrophysiological criteria (Rager 1976b). This fits into our axonal flow model (Sect. 6.2) which implies that already the beginning of the axonal arborization is critical for the further maturation of the cell. However, there is some evidence from our Golgi preparations that the growing tips of retinal axons come into close proximity to tectal dendrites from incubation day 8 onward (Rager and von Oeynhausen 1979). Thus it might be possible that the initial axonal arborization already entails some kind of molecular interaction between afferent fibres and tectal target cells even at this stage, leading to some sort of selective stabilization (Changeux et al. 1973) of the surviving ganglion cells. Non-stabilized cells will die. This is expressed in the concept of a competent cell introduced above.

Competition and degeneration seem to be simple and powerful mechanisms working on the formation of connections in the brain. They lead to systems-matching of the retina and the optic tectum which are certainly not matched in size initially. In addition, if cells influence pre-synaptic elements, a functional selection can be imagined which to a certain extent is able to eliminate "unsuitable" connections in a microenvironment. Such mechanisms could be of great importance in the development of the retinotectal projection of the chicken, because up to now no evidence has been found showing a major shift or transposition of retinotectal synapses which could finally lead to a precise retinotopy. Again it appears that rather non-specific processes are sufficient to build up highly specific connections.

7 Summary

Retinal ganglion cells are first generated in the centre of the retina. During migration towards the inner limiting membrane each of these cells forms an axon. Axons join together, form fascicles and grow to the choroid fissure by the shortest possible route. The development of the cell goes through bipolar, unipolar and multipolar stages. Ganglion cells do not form their mature dendrites until their axons have reached their termination sites. The axons of the earliest-generated ganglion cells reach the central tectal area on incubation day 8. They immediately invade the cortical plate and begin to ramify. The ramification of axons progresses rapidly. Synapses could first be detected with the electron microscope in the central tectal area on incubation day 10. They probably originate from retinal fibres.

At early stages of development the optic nerve consists of neuroepithelial cells, growth cones and thin axons. From incubation day 10 onwards large bundles of uniformly thin axons dominate. On incubation day 14 some axons have already increased in diameter and glial cells form fine processes which begin to envelop the thicker axons. Myelination begins in the central core of the nerve one day later. At hatching 6% of the fibres are already myelinated. However, most of the myelination process takes place only after hatching. Three months after hatching 94% of fibres are myelinated. All axons larger than about 0.75 μm, are myelinated, thus myelination appears to be a function of axon diameter.

Ganglion cell axons are able to transmit action potentials at least from incubation day 8 onwards which is when the first ganglion cells change over from their multipolar to their mature stage. Synaptic transmission can first be detected on incubation day 10 with the aid of a current-source-density analysis. At this time central retinal axons ramify extensively in the central tectal region and synapses can be found in the upper tectal layers for the first time. Thus, morphological and physiological aspects of synaptogenesis are highly correlated.

The formation of the retinotectal map and the final maturation of ganglion cells have been studied from the point of view of developmental kinetics and dynamics. The generation of ganglion cells spreads from the centre of the retina to the periphery. Thus, each ganglion cell can be characterized by the space-time coordinates of its generation. Ganglion cell axons are arranged in an orderly manner in the retina and in the visual pathway and they tend to maintain the neighbourhood relations of their perikarya throughout development. The degree of order found in retinal fibres may be sufficient to explain the order measured in morphological and physiological experiments, which is limited not only by the arrangement of fibres in the optic nerve, but also by the overlap of axonal terminal arbors. Like ganglion cells, tectal neurones mature first in the central tectal area and only later in peripheral tectal regions. When axons of the first generated ganglion cells arrive at the rostral pole of the tectum (incubation day 6), the tectum is very immature and its cortical plate is not yet formed. The axons apparently continue to grow on the tectal surface until they meet dendrites mature enough to receive contacts which is on incubation day 8 and at the central tectal area. Axons do not "wait" for two days, but immediately start to ramify and to invade the cortical plate at that site. The synchrony between the arrival of central retinal axons and the maturation of central tectal neurones would be sufficient to determine the origin of the retinotectal map. The orientation of the mapping onto peripheral tectal regions could be due to the organization in space and time of the arriving fibre tract and to the fact that the maturation of tectal cells progresses continuously from the centre to the periphery. All these factors taken together would allow one to formulate a consistent hypothesis for the formation of the retinotectal map without invoking the existence of chemical markers.

The final maturation of ganglion cells seems to depend on critical events occurring in the axonal periphery. If the formation of the final dendrite is a criterion for the maturation of ganglion cells, this step in maturation apparently only occurs when axons have arrived at their termination area. When an axon begins to ramify, its diameter increases as a function of the number of terminal branches. Thus, it is postulated that each axon needs a minimum number of branches or synaptic contacts for the final maturation and survival of the cell. This means that a minimum number of tectal cells would be required per axon.

This being the case, if there are more ganglion cell axons arriving at the tectum than available termination sites, supernumerary cells would die. In fact, from incubation days 8 and 9 onwards, when the first retinal fibres have arrived at the optic tectum and started branching, ganglion cells do begin to degenerate in the retina. Degeneration, like generation, spreads from the centre to the periphery of the retina and reflects the pattern of ingrowth of fibres into the optic tectum. The total number of ganglion cells decreases from 4 to 2.4 million which is a reduction of 40%. The final value is reached on incubation day 18, i.e. when photoreceptors start to respond to light stimuli.

A mathematical model has been developed, based on the propositions that fibres compete for a minimum number of termination sites and that supernumarary cells degenerate. It turns out that the measured values can be interpreted by the model very well. Thus it can be concluded that the two systems, retina and optic tectum, could be matched in size by degeneration. Finally, it has to be emphasized that in the chicken the whole retinotectal projection is laid down before light stimuli can exert any influence.

References

Albus K (1975) A quantitative study of the projection area of the central and the paracentral visual field in area 17 of the cat. I. The precision of the topography. Exp Brain Res 24:159–179

Ansinn KD, Oehlschlägel HK, Michael D (1969) Fußpunkt-Latenzen im Tectum opticum von Hühnerembryonen. Pfluegers Arch 307:151–152

Barron DH (1943) The early development of the motor cells and columns in the spinal cord of the sheep. J Comp Neurol 78:1–27

Barron DH (1946) Observations on the early differentiation of the motor neuroblasts in the spinal cord of the chick. J Comp Neurol 85:149–169

Binggeli RL, Paule WJ (1969) The pigeon retina: quantitative aspects of the optic nerve and ganglion cell layer. J Comp Neurol 137:1–18

Blechschmidt E (1961) Die vorgeburtlichen Entwicklungsstadien des Menschen. Karger, Basel

Blechschmidt E (1968) Vom Ei zum Embryo. Deutsche Verlags-Anstalt, Stuttgart

Blechschmidt E (1974) Humanembryologie. Prinzipien und Grundbegriffe. Hippokrates, Stuttgart

Blechschmidt E (1977) The programming of afferent and efferent nervous fibers in man. Arch Psychiatr Nervenkr 224:259–272

Blechschmidt E, Gasser RF (1978) Biokinetics and biodynamics of human differentiation. Principles and applications. Thomas, Springfield

Blozovski D, Blozovski M (1968) Développement comparé de l'électrorétinogramme et des potentiels évoqués visuels du toit optique du cervelet et du télencéphale chez le poussin. J Physiol (Paris) 60:33–50

Bodian D (1937) The staining of paraffin sections of nervous tissues with activated protargol. The role of fixatives. Anat Rec 69:153–162

Boycott BB, Dowling JE (1969) Organization of the primate retina: light microscopy. Philos Trans R Soc Lond (Biol) 255:109–184

Boycott BB, Wässle H (1974) The morphological types of ganglion cells of the domestic cat's retina. J. Physiol (Lond) 240:397–419

Buren JM van (1963) The retinal ganglion cell layer. Thomas, Springfield

Butler R (1971) Very rapid selective silver (Golgi) impregnations and embedding of invertebrate nervous tissue. Brain Res 33:540–544

Changeux JP, Courrège Ph, Danchin A (1973) A theory of the epigenesis of neuronal networks by selective stabilization of synapses. Proc Natl Acad Sci USA 70:2974–2978

Chung SH, Bliss TVB, Keating MJ (1974) The synaptic organization of optic afferents in the amphibian tectum. Proc R Soc Lond (Biol) 187:421–447

Chu-Wang IW, Oppenheim RW (1978) Cell death of motoneurons in the chick embryo spinal cord. I. A light and electron microscopic study of naturally occurring and induced cell loss during development. J Comp Neurol 177:33–58

Clarke PGH, Cowan WM (1976) The development of the isthmo-optic tract in the chick, with special reference to the occurrence and correction of developmental errors in the location and connections of isthmo-optic neurons. J Comp Neurol 167:143–164

Clarke PGH, Whitteridge D (1976) The projection of the retina, including the "red area". on to the optic tectum of the pigeon. Q J Exp Physiol 61:351–358

Colonnier M (1964) The tangential organisation of the visual cortex. J Anat 98:327–344

Constantine-Paton M, Capranica R (1976a) Axonal guidance of developing optic nerves in the frog. I. Anatomy of the projection from transplanted eye primordia. J Comp Neurol 170:17–32

Constantine-Paton M, Capranica R (1976b) Axonal guidance of developing optic nerves in the frog. II. Electrophysiological studies of the projection from transplanted eye primordia. J Comp Neurol 170:33–52

Constantine-Paton M, Law MI (1978) Eye-specific termination bands in tecta of three-eyed frogs. Science 202:639–641

Cook JE, Horder IJ (1977) The multiple factors determining retinotopic order in the growth of optic fibres into the optic tectum. Philos Trans R Soc Lond (Biol) 278:261–276

Cowan WM (1973) Neuronal death as a regulative mechanism in the control of cell number in the nervous system. In: Rockstein M (ed) Development and aging in the nervous system. Academic Press, New York, pp 19–41

Cowan WM, Martin AH, Wenger E (1968) Mitotic patterns in the optic tectum of the chick during normal development and after early removal of the optic vesicle. J Exp Zool 169:71–92

Creutzfeldt OD, Sakmann B, Scheich H, Korn A (1970) Sensitivity distribution and spatial summation within receptive-field center of retinal on-center ganglion cells and transfer function of the retina. J Neurophysiol 33:654–671

Creutzfeldt OD, Innocenti GM, Brooks D (1974) Vertical organisation in the visual cortex (Area 17) in the cat. Exp Brain Res 21:315–336

Crossland WJ, Cowan WM, Rogers LA, Kelly JP (1974) The specification of the retino-tectal projection in the chick. J Comp Neurol 155:127–164

Crossland WJ, Cowan WM, Rogers LA (1975) Studies on the development of the chick optic tectum. IV. An autoradiographic study of the development of retinotectal connections. Brain Res 91:1–23

DeLong GR, Coulombre AJ (1965) Development of the retinotectal topographic projection in the chick embryo. Exp Neurol 13:351–363

Erlanger J, Gasser HS (1937) Electric signs of nervous activity. University of Philadelphia Press, Pennsylvania

Fischer B (1973) Overlap of receptive field centers and representation of the visual field in the cat's optic tract. Vision Res 13:2113–2120

Galifret Y (1968) Les diverses aires fonctionelles de la rétine du pigeon. Z Zellforsch 86:535–545

Garcia-Austt E, Patetta-Queirolo MA (1961) Electroretinogram of the chick embryo. I. Onset and development. Acta Neurol Lat Am 7:179–189

Gaze RM, Keating MJ (1972) The visual system and "neuronal specificity." Nature 237:375–378

Giorgi PP, Van der Loos H (1978) Directed growth of optic fibres in amphibia. Neurosci Lett [Suppl] 1:390

Goldberg S (1972) Silver staining, featuring rapid reduction, for whole mounts of retina and optic pathways in chick embryos. Stain Technol 47:65–69

Goldberg S (1974) Studies on the mechanics of development of the visual pathways in the chick embryo. Dev Biol 36:24–43

Goldberg S (1976) Polarization of avian retina. Ocular transplantation studies. J Comp Neurol 168:379–392

Goldberg S, Coulombre AJ (1972) Topographical development of the ganglion cell fiber layer in the chick retina. A whole mount study. J Comp Neurol 146:507–518

Hally AD (1964) A counting method for measuring the volumes of tissue components in microscopical sections. QJ Microsc Sci 105:503–517

Hamburger V, Hamilton HL (1951) A series of normal stages in the development of the chick embryo. J Morphol 88:49–92

Hamdi FA, Whitteridge D (1954) The representation of the retina on the optic tectum of the pigeon. QJ Exp Physiol 39:111–118

Hartley HO (1948) The estimation of non-linear parameters by "internal least squares". Biometrika 35:32–45.

Hartley HO (1961) The modified Gauss-Newton method for the fitting of nonlinear regression functions by least squares. Technometrics 3:269–280

Hebel R (1976) Distribution of retinal ganglion cells in five mammalian species (Pig, sheep, ox, horse, dog). Anat Embryol (Berl) 150:45–51

Henschen SE (1910) Zentrale Sehstörungen. In: Lewandowski MH (ed) Handbuch der Neurologie. Springer: Berlin Heidelberg New York, pp 891–918

Hilliard JE (1968) Measurement of volume in volume. In: Hoff RT de, Rhines FN (eds) Quantitative microscopy. McGraw-Hill, New York, pp 45–76

Hilliard JE, Cahn JW (1961) An evaluation of procedures in quantitative metallography for volume-fraction analysis. Trans Metal Soc AIME 221:344–352

Hinds JW, Hinds PL (1974) Early ganglion cell differentiation in the mouse retina: an electron microscopic analysis utilizing serial sections. Dev Biol 37:381–416

His W (1874) Unsere Körperform und das physiologische Problem ihrer Entstehung. Vogel, Leipzig

His W (1901) Das Prinzip der organbildenden Keimbezirke und die Verwandtschaften der Gewebe. Arch Anat Physiol 307–337

Holden AL (1968a) The field potential profile during activation of the avian optic tectum. J Physiol (Lond) 194:75–90

Holden AL (1968b) Types of unitary response and correlation with the field potential profile during activation of the avian optic tectum. J Physiol (Lond) 194:91–104

Hope RA, Hammand BJ, Gaze RM (1976) The arrow model: retinotectal specificity and map formation in the goldfish visual system. Proc R Soc Lond (Biol) 194:447–466

Horder TJ, Mashkas A, Pilgrim AJ (1979) A method for the determination of fibre organization within the visual pathways of higher vertebrates. J Physiol (Lond) 296:8P

Horder TJ, Mashkas A, Webb JN (1979) Morphogenetic forces in the development of the avian retina of possible significance for the polarity of central visual projections. J Physiol (Lond) 291:12–13P

Hubel DH, Wiesel TN (1974) Uniformity of monkey striate cortex: a parallel relationship between field size, scatter, and magnification factor. J Comp Neurol 158:295–306

Hughes AF (1953) The growth of embryonic neurites. A study on cultures of chick neural tissue. J Anat 87:150–162

Hughes A (1975) A quantitative analysis of the cat retinal ganglion cell topography. J Comp Neurol 163:107–128

Hughes A, Wässle H (1976) The cat optic nerve: Fibre total count and diameter spectrum. J Comp Neurol 169:171–184

Hughes WF, La Velle A (1974) On the synaptogenic sequence in the chick retina. Anat Rec 179:297–302

Hunt RK, Jacobson M (1974) Neuronal specificity revisited. Curr Top Dev Biol 8:203–259

Ito S, Winchester RJ (1963) The fine structure of the gastric mucosa in the bat. J Cell Biol 16:541–577

Kahn AJ (1973) Ganglion cell formation in the chick neural retina. Brain Res. 63:285–290

Kahn AJ (1974) An autoradiographic analysis of the time of appearance of neurons in the developing chick neural retina. Dev Biol 38:30–40

Katz MJ, Lasek RJ (1979) Substrate pathways which guide growing axons in *Xenopus* embryos. J Comp Neurol 183:817–832

Kretschmann HJ, Wingert F (1971) Computeranwendungen bei Wachstumsproblemen in Biologie und Medizin. Springer, Berlin Heidelberg New York

Kreyszig E (1968) Statistische Methoden und ihre Anwendung. Vandenhoeck and Ruprecht, Göttingen

Kuffler SW, Nicholls JG (1966) The physiology of neuroglial cells. Ergeb Physiol 57:1–90

LaVail JH, Cowan WM (1971a) The development of the chick optic tectum. I. Normal morphology and cytoarchitectonic development. Brain Res 28:391–419

LaVail JH, Cowan WM (1971b) The development of the chick optic tectum. II. Autoradiographic studies. Brain Res 28:421–441

Leghissa S (1957) Il differenziamento ontogenetico e istogenetico del tetto ottico nell' embrione di pollo. Arch Sci Biol (Bologna) 41:601–628

Leibnitz L (1967) Die Veränderung von Gewicht, Volumen und spezifischem Gewicht des Rattengehirns nach Fixierung, Dehydrierung und Aufhellung. J Hirnforsch 9:97–104

Lund RD, Bunt AH (1976) Prenatal development of central optic pathways in albino rats. J Comp Neurol 165:247–264

Malsburg C von der, Willshaw DJ (1977) How to label nerve cells so that they can interconnect in an ordered fashion. Proc Natl Acad Sci USA 74:5176–5178

McGill JI, Powell TPS, Cowan WM (1966) The retinal representation upon the optic tectum and isthmo-optic nucleus in the pigeon. J Anat 100:5–33

Miles FA (1972) Centrifugal control of the avian retina. I. Receptive field properties of retinal ganglion cells. Brain Res 48:65–92

Mitzdorf U, Singer W (1977) Laminar segregation of afferents to lateral geniculate nucleus of the cat: an analysis of current source density. J Neurophysiol 40:1227–1244

Mollenhauer HH (1964) Plastic embedding mixtures for use in electron microscopy. Stain Technol 39:111–114

Morest DK (1970) The pattern of neurogenesis in the retina of the rat. Z Anat Entwicklungsgesch 131:45–67

Orkand RK, Nicholls JG, Kuffler SW (1966) Effect of nerve impulses on the membrane potential of glial cells in the central nervous system of amphibia. J Neurophysiol 29:788–806

Patterson JL Jr, Goetz RH, Doyle JT, Warren JV, Gauer OH, Detweiler DK, Said SI, Hoernicke H, McGregor M, Keen EN, Smith MH Jr, Hardie EL, Reynolds M, Flatt WP, Waldo DR (1965) Cardiorespiratory dynamics in the ox and giraffe, with comparative observations on man and other mammals. Ann NY Acad Sci 127:393–413

Peil, J (1970) Ein Verfahren zur nichtlinearen Approximation und seine Anwendung auf verschiedene naturwissenschaftliche, technische und medizinische Probleme. Nova Acta Leopold (NS) 35:195

Peters JJ, Vonderahe AR, Powers TH (1958) Electrical studies of functional development of the eye and optic lobes in the chick embryo. J Exp Zool 139:459–468

Pilar G, Landmesser L (1976) Ultrastructural differences during embryonic cell death in normal and peripherally deprived ciliary ganglia. J Cell Biol 68:339–356

Pisareva NL (1965) Bioelectrical responses from midbrain tectum evoked by photic stimuli in chick embryo. Zh Evo Biokhim Fiziol 1:175–182

Polyak SL (1957) The vertebrate visual system. Univ of Chicago Press: Chicago

Popper KR (1959) The logic of scientific discovery. Hutchinson: London

Puelles L, Bendala MC (1978) Differentiation of neuroblasts in the chick optic tectum up to eight days of incubation: A Golgi study. Neuroscience 3:307–325

Rager G (1976a) Morphogenesis and physiogenesis of the retino-tectal connection in the chicken. I. The retinal ganglion cells and their axons. Proc R Soc Lond. (Biol) 192:331–352

Rager G (1976b) Morphogenesis and physiogenesis of the retino-tectal connection in the chicken. II. The retino-tectal synapses. Proc R Soc Lond. (Biol) 192:353–370

Rager G (1977) Development of the electroretinogram in the chick embryo. A correlation of structure and function. Acta Anat 99:305

Rager G (1978) Systems-matching by degeneration. II. Interpretation of the generation and degeneration of retinal ganglion cells by a mathematical model. Exp Brain Res 33:79–90

Rager G (1979) The cellular origin of the b-wave in the electroretinogram. A developmental approach. J Comp Neurol 188:225–244

Rager G (1980) Die Ontogenese der retinotopen Projektion. Beobachtung und Reflexion. Naturwissenschaften 67:280–287

Rager G, Kreische R (1978) A current-source-density analysis in the developing optic tectum of the chicken. Neurosci Lett (Suppl) 1:394

Rager G, Oeynhausen B von (1979) Ingrowth and ramification of retinal fibres in the developing optic tectum of the chick embryo. Exp Brain Res 35:213–227

Rager G, Rager U (1978) Systems-matching by degeneration. I. A quantitative electromicroscopic study of the generation and degeneration of retinal ganglion cells in the chicken. Exp Brain Res 33:65–78

Rager G, Lausmann S, Gallyas F (1979) An improved silver stain for developing nervous tissue. Stain Technol 54:193–200

Rager G, Behrmann A, Kreische R (1980) Synaptogenesis in the optic tectum of the chick as revealed by a current-source-density analysis. In press

Rager G, Tanaka M, Nowakowski RS The arrangement of retinal fibres in the visual pathway of the chick as revealed by the injection of HRP. In preparation

Raisman G (1973) An ultrastructural study of the effects of hypophysectomy on the supraoptic nucleus of the rat. J Comp Neurol 147:181–208

Rakic P (1977) Prenatal development of the visual system in rhesus monkey. Philos Trans R Soc Lond (Biol) 278:245–260

Ramón y Cajal S (1893) La rétine des vértebrés. Cellule 9:17–257

Ramón y Cajal S (1909) Histologie du système nerveux de l'homme et des vertébrés, vol I. New edition 1972, Instituto Ramón y Cajal, Madrid

Ramón y Cajal S (1911) Histologie du système nerveux de l'homme et des vertébrés, vol II. New edition 1972, Instituto Ramón y Cajal, Madrid

Ramón y Cajal S (1913) Études sur la dégénération et régénération du système nerveux. Madrid 1913–1914

Reynolds ES (1963) The use of lead citrate at high pH as an electron-opaque stain in electron microscopy. J Cell Biol 17:208–212

Romeis B (1968) Mikroskopische Technik. Oldenbourg, München

Scharf JH (1969) Zum Körperlängenwachstumsgesetz der menschlichen Leibesfrucht. Acta Anat. (Basel) 73:10–18

Scharf JH (1970) Innere Regressionsrechnung mit nichtlinearen Differentialgleichungen. Biometr. Z 12:228–241

Scharf JH (1971) Differentialgleichungen in der funktionellen Morphologie. Gegenbaurs Morphol Jahrb 117:3–38

Scharf JH, Hoffmann M (1971) Zwei bequeme Algorithmen zur Anpassung von Sigmoidkurven an Meßwerte nach dem Prinzip der kleinsten Fehlerquadrate. Nova Acta Leopold (NS) 36/202:11–130

Scharf JH, Peil J (1975) Ein Algorithmus zur Wertebestimmung der Parameter in der Gompertz-schen Wachstumsfunktion. Gegenbaurs morphol Jahrb 121:389–420

Schmidt JT (1977) Retinal fibres alter tectal positional markers during the expansion of the half retinal projection in goldfish. J Comp Neurol 177:279–300

Scholes JH (1979) Nerve fibre topography in the retinal projection to the tectum. Nature 278:620–624

Sedláček I (1969) The responses of optic evoked potentials to strychnine in chick embryos. Exp Brain Res 9:357–364

Sheffield JB, Fischman DA (1970) Intercellular junctions in the developing neural retina of the chick embryo. Z Zellforsch 104:405–418

Silver J, Robb RM (1979) Studies on the development of the eye cup and optic nerve in normal mice and in mutants with congenital optic nerve aplasia. Develop Biol 68:175–190

Silver J, Sidman RL (1980) A mechanism for the guidance and topographic patterning of retinal ganglion cell axons. J Comp Neurol 189:101–111

Singer M, Nordlander RH, Egar M (1979) Axonal guidance during embryogenesis and regeneration in the spinal cord of the newt: the blueprint hypothesis of neuronal pathway patterning. J Comp Neurol 185:1–22

Sperry RW (1943) Visuomotor co-ordination in the newt (Triturus viridescens) after regeneration of the optic nerve. J Comp Neurol 79:33–55

Sperry RW (1963) Chemoaffinity in the orderly growth of nerve fiber patterns and connections. Proc Natl Acad Sci USA 50:703–710

Stone J (1965) A quantitative analysis of the distribution of ganglion cells in the cat's retina. J Comp Neurol 124:337–352

Stone J, Freeman JA (1971) Synaptic organisation of the pigeon's optic tectum: a Golgi and current source-density analysis. Brain Res 27:203–221

Stone J, Holländer H (1971) Optic nerve axon diameters measured in the cat retina: some functional considerations. Exp Brain Res 13:498–503

Suburo A, Carri N, Adler R (1979) The environment of axonal migration in the developing chick retina: a scanning electron microscopic (SEM) study. J Comp Neurol 184:519–536

Sur M, Merzenich MM, Kaas JH (1979) Magnification, receptive field area and "hypercolumn" size in somatosensory cortex of the owl monkey. Soc Neurosci Abstr 5:713

Szentágothai J, Székely G (1956) Zum Problem der Kreuzung von Nervenbahnen. Acta Biol Acad Sci Hung 6:215–229

Thom R (1975) Structural stability and morphogenesis. An outline of a general theory of models. Benjamin, Reading

Vaney DI, Hughes A (1976) The rabbit optic nerve: Fibre diameter spectrum, fibre count, and comparison with a retinal ganglion cell count. J Comp Neurol 170:241–252

Vaughn JE, Peters A (1968) A third neuroglial cell type. J Comp Neurol 133:269–288

Waerden BL van der (1971) Mathematische Statistik. Springer, Berlin Heidelberg New York

Weiss P (1928) Eine neue Theorie der Nervenfunktion. Nicht durch gesonderte Bahnen, sondern durch spezifische Formen der Erregung schaltet das Nervensystem mit den Muskeln. Naturwissenschaften 16:626–636

Weiss P (1929) Erzwingung elementarer Strukturverschiedenheiten am in vitro wachsenden Gewebe. Arch Entwicklungsmech Organ 16:438–554

Weiss P (1934) In vitro experiments on the factors determining the course of the out-growing nerve fiber. J Exp Zool 68:393–448

Weiss P (1936) Selectivity controlling the central-peripheral relations in the nervous system. Biol Rev 11:494–531

Weiss P (1941) Nerve Patterns. The mechanics of nerve growth. Growth Symp 5:163–203

Weiss P (1955) Nervous system. In: Willier B, Weiss P, Hamburger V (eds) Analysis of development. Saunders, Philadelphia, pp 346–401

Willshaw DJ, Malsburg Ch von der (1979) How patterned neural connections can be set up by self-organization. Philos Trans R Soc Lond (Biol) 287:203–243

Wingert F (1969) Biometrische Analyse der Wachstumsfunktionen von Hirnteilen und Körpergewicht der Albinomaus. J Hirnforsch 11:133–197

Witkovsky P (1963) An ontogenetic study of retinal function in the chick. Vision Res 3:341–355

Zilles KJ (1978) Ontogenesis of the visual system. Adv Anat Embryol Cell Biol 54:3

Zilles K, Schleicher A, Kretschmann HJ, Wingert F (1976) Semiautomatic morphometric analysis of the nucleolar development in the Nucl. N. oculomotorii of *Tupaia belangeri* during ontogenesis. Anat Embryol (Berl) 149:15–28

Zurmühl R (1965) Praktische Mathematik für Ingenieure und Physiker. Springer, Berlin Heidelberg New York

Subject Index

Other Reviews of Interest in this Series

Volume 56

Kaissling, B., Kriz, W.:
Structual Analysis of the Rabbit Kidney.
47 figures. VIII, 123 pages. 1979.
ISBN 3-540-09145-9

Volume 57

Niimi, K., Matsuoka, H.:
Thalamocortical Organization of the
Auditory System in the Cat Studied by
Retrograde Axonal Transport of Horse-
radish Peroxidase.
30 figures. X, 56 pages. 1979.
ISBN 3-540-09449-0

Volume 58

Verwoerd, C.D.A., van Oostrom, C.G.:
Cephalic Neural Crest and Placodes.
41 figures. VI, 75 pages. 1979.
ISBN 3-540-09608-6

Volume 59

Bär, T.: The Vascular System of the
Cerebral Cortex.
33 figures. VI, 60 pages. 1980.
ISBN 3-540-09652-3

Volume 60

Hildebrand, R.: Nuclear Volume
and Cellular Metabolism.
12 figures. VII, 54 pages. 1980.
ISBN 3-540-09796-1

Volume 61

Korr, H.: Proliferation of Different Cell
Types in the Brain.
21 figures. VII, 72 pages. 1980.
ISBN 3-540-09899-2

Volume 62

Brown Gould, B.: Organization
of Afferents from the Brain Stem
Nuclei to the Cerebellar Cortex
in the Cat.
10 figures. VIII, 90 pages. 1980.
ISBN 3-540-09960-3

Springer-Verlag Berlin Heidelberg New York